Review of New York State Low-Level Radioactive Waste Siting Process

Committee to Review New York State's Siting and Methodology
Selection for Low-Level Radioactive Waste Disposal

Board on Radioactive Waste Management
Commission on Geosciences, Environment, and Resources
National Research Council

NATIONAL ACADEMY PRESS
Washington, D.C. 1996

NOTICE: The project that is the subject of this report was approved by the Governing Board of the National Research Council, whose members are drawn from the councils of the National Academy of Sciences, the National Academy of Engineering, and the Institute of Medicine. The members of the committee responsible for the report were chosen for their special competences and with regard for appropriate balance.

This report has been reviewed by a group other than the authors according to procedures approved by the Report Review Committee consisting of members of the National Academy of Sciences, the National Academy of Engineering, and the Institute of Medicine.

Support for this study was provided by the New York State Department of Health under Contract No. C-010778. All opinions, findings, conclusions, or recommendations expressed herein are those of the author(s) and do not necessarily reflect the views of the New York State Department of Health.

Library of Congress Catalog Card Number 96-69080
International Standard Book Number 0-309-05539-3

Additional copies of this report are available from:

National Academy Press
2101 Constitution Avenue, NW
Box 285
Washington, DC 20055
800-624-6242
202-334-3313 (in the Washington Metropolitan Area)

Cover art entitled "Variations on a Theme," ©1994, by Edith Socolow. An artist of experience and many credits, Edith Socolow creates images of land and sea. Her landscapes are exaggerated or partially obscured and left for the observer to interpret. Ms. Socolow resides in Gloucester, Massachusetts.

Printed in the United States of America

COMMITTEE TO REVIEW NEW YORK STATE'S SITING AND METHODOLOGY SELECTION FOR LOW-LEVEL RADIOACTIVE WASTE DISPOSAL

SUSAN WILTSHIRE, *Chair*, JK Research Associates Inc., Hamilton, Massachusetts
ROBERT J. AHRENS, National Soil Survey Center/U.S. Department of Agriculture, National Resources Conservation Service, Lincoln, Nebraska
GLORIA ANDERSON, League of Women Voters of California, Crestline
CHARLES BASKERVILLE, Central Connecticut State University, New Britain
RANDY L. BASSETT, University of Arizona, Tucson
LYNDA L. BROTHERS, Davis Wright Tremaine, Seattle, Washington
HALINA BROWN, Clark University, Worchester, Massachusetts
GAIL CEDERBERG, Consultant, Colts Neck, New Jersey
JOHN CROES, Science Applications International Corporation, Oak Ridge, Tennessee
WILLIAM P. DORNSIFE, Pennsylvania Department of Environmental Protection, Harrisburg
JOHN E. EBEL, Weston Observatory, Boston College, Weston, Massachusetts
WILLIAM FREUDENBURG, University of Wisconsin, Madison
ROBERT D. HATCHER, JR., The University of Tennessee and Oak Ridge National Laboratory, Knoxville and Oak Ridge
CAROL HORNIBROOK,* Electric Power Research Institute, Palo Alto, California
JANET JOHNSON, Shepherd Miller, Inc., Fort Collins, Colorado
LINDA L. LEHMAN, L. Lehman & Associates, Inc., Prior Lake, Minnesota
ROBERT MEYER, Keystone Scientific, Inc., Fort Collins, Colorado
DELLA ROY, Pennsylvania State University, University Park
MIKLOS D.G. SALAMON,† Colorado School of Mines, Golden
LEONARD C. SLOSKY, Slosky & Company, Inc., Denver, Colorado
ARTHUR A. SOCOLOW, Pennsylvania Geological Survey (retired), Gloucester, Massachusetts

NRC Staff

INA B. ALTERMAN, Study Director (through 6/95)
KEVIN D. CROWLEY, Study Director (since 6/95)
CHARLES MEADE, Senior Staff Officer
REBECCA BURKA, Senior Project Assistant
ELIZABETH M. LANDRIGAN, Project Assistant
ERIKA L. WILLIAMS, Project Assistant
SCOTT A. HASSELL, Intern

Consultant

CINDY M. MONACO

---

*Resigned due to other professional commitments.
†Resigned due to change in scope of report.

BOARD ON RADIOACTIVE WASTE MANAGEMENT

MICHAEL C. KAVANAUGH, *Chair,* ENVIRON Corporation, Emeryville, California
B. JOHN GARRICK, *Vice-Chair,* PLG, Inc., Newport Beach, California
JOHN F. AHEARNE, Sigma Xi, The Scientific Research Society, and Duke University, Research Triangle Park and Durham, North Carolina
JEAN M. BAHR, University of Wisconsin, Madison
SOL BURSTEIN, Wisconsin Electric Power, Milwaukee (retired)
ANDREW P. CAPUTO, Natural Resources Defense Council, Washington, D.C.
MELVIN W. CARTER, Georgia Institute of Technology, Atlanta (emeritus)
PAUL P. CRAIG, University of California, Davis (emeritus)
MARY R. ENGLISH, The University of Tennessee, Knoxville
ROBERT D. HATCHER, JR., The University of Tennessee and Oak Ridge National Laboratory, Knoxville and Oak Ridge
DARLEANE C. HOFFMAN, Lawrence Berkeley Laboratory, Berkeley, California
JAMES H. JOHNSON, JR., Howard University, Washington, D.C.
CHARLES McCOMBIE, NAGRA, Wettingen, Switzerland
ROBERT MEYER, Keystone Scientific, Inc., Fort Collins, Colorado
PRISCILLA P. NELSON, University of Texas, Austin
D. KIRK NORDSTROM, U.S. Geological Survey, Boulder, Colorado
D. WARNER NORTH, Decision Focus, Inc., Mountain View, California
PAUL SLOVIC, Decision Research, Eugene, Oregon
BENJAMIN L. SMITH, Independent Consultant, Columbia, Tennessee

NRC Staff

KEVIN D. CROWLEY, Director (from 6/96)
CARL A. ANDERSON, Director (through 5/96)
ROBERT S. ANDREWS, Senior Staff Officer
KARYANIL T. THOMAS, Senior Staff Officer
THOMAS E. KIESS, Staff Officer
SUSAN B. MOCKLER, Research Associate
LISA J. CLENDENING, Administrative Assistant
ROBIN L. ALLEN, Senior Project Assistant
REBECCA BURKA, Senior Project Assistant
DENNIS L. DUPREE, Senior Project Assistant
PATRICIA A. JONES, Project Assistant
ANGELA R. TAYLOR, Project Assistant
ERIKA L. WILLIAMS, Project Assistant
JOSHUA A. CHAMOT, Intern

COMMISSION ON GEOSCIENCES, ENVIRONMENT, AND RESOURCES

The National Academy of Sciences is a private, nonprofit, self-perpetuating society of distinguished scholars engaged in scientific and engineering research, dedicated to the furtherance of science and technology and to their use for the general welfare. Upon the authority of the charter granted to it by the Congress in 1863, the Academy has a mandate that requires it to advise the federal government on scientific and technical matters. Dr. Bruce Alberts is president of the National Academy of Sciences.

The National Academy of Engineering was established in 1964, under the charter of the National Academy of Sciences, as a parallel organization of outstanding engineers. It is autonomous in its administration and in the selection of its members, sharing with the National Academy of Sciences the responsibility for advising the federal government. The National Academy of Engineering also sponsors engineering programs aimed at meeting national needs, encourages education and research, and recognizes the superior achievements of engineers. Dr. William A. Wulf is interim president of the National Academy of Engineering.

The Institute of Medicine was established in 1970 by the National Academy of Sciences to secure the services of eminent members of appropriate professions in the examination of policy matters pertaining to the health of the public. The Institute acts under the responsibility given to the National Academy of Sciences by its congressional charter to be an adviser to the federal government and, upon its own initiative, to identify issues of medical care, research, and education. Dr. Kenneth Shine is president of the Institute of Medicine.

The National Research Council was organized by the National Academy of Sciences in 1916 to associate the broad community of science and technology with the Academy's purposes of furthering knowledge and of advising the federal government. Functioning in accordance with general policies determined by the Academy, the Council has become the principal operating agency of both the National Academy of Sciences and the National Academy of Engineering in providing services to the government, the public, and the scientific and engineering communities. The Council is administered jointly by both Academies and the Institute of Medicine. Dr. Bruce Alberts and Dr. William A. Wulf are the chairman and interim vice-chairman, respectively, of the National Research Council.

Preface

In 1986, the New York State Legislature enacted the New York State Low-Level Radioactive Waste Management Act. Among its provisions was the establishment of a commission charged with identifying a site in the state for a low-level waste disposal facility. To carry out its work, the Sitting Commission developed a stepwise technical screening process based on consideration of factors such as surface and ground water hydrology, geologic properties, demographic issues, land use patterns, and socioeconomic concerns. The screening process was designed to select sites for more detailed analysis of their suitability for hosting a potential repository.

By June 1990 it was clear that the siting process had reached an impasse due to local opposition. Consequently, the New York State Legislature made broad changes to the 1986 State Act calling for, among other provisions, the creation of an independent technical and scientific panel to review the work of the Siting Commission, as described in Chapter 1 of this report.

The New York State Energy Research and Development Agency first requested the National Research Council (NRC) to review New York State's siting process in March of 1990, prior to the above-mentioned legislative changes. A proposal was submitted by the NRC at that time. When the July 1990 legislation explicitly defined the panel's role, New York State developed a detailed scope of work for the panel that extended beyond the scope of the first NRC proposal. The new legislation also designated the New York State Department of Health as the agency responsible for supporting the independent scientific and technical review panel. In June of 1992 the Department of Health invited the NRC to submit a revised proposal. In April 1993 the Board on Radioactive Waste Management was contracted by the Department of Health to undertake the review; and subsequently, the Committee to Review New York State's Siting and Methodology Selection for Low-Level Radioactive Waste Disposal (hereafter referred to as the "committee") was formed.

The committee was asked by the Department of Health to review the scientific, technical, and procedural approaches used by the Siting Commission. The committee was directed to review and comment on the criteria, methodology, procedures, and decision process used by the Siting

Commission to select potential low-level radioactive waste disposal sites and was asked to address eight specific technical questions (see Chapter 8). A complete statement of task is given in Appendix B of this report.

The committee held its first meeting in January 1994, entering a polarized situation in which it seemed that all sides expected the committee's report to justify their positions. During our review, it became increasingly obvious that the process we were reviewing would not continue in its current form. Indeed, the New York State Legislature voted to end the activities of the Siting Commission in 1995, leaving the committee to wonder whether its work had any relevance beyond providing criticism or justification for past actions. There are, however, many lessons from our review that could help others responsible for siting potentially controversial facilities. This report, therefore, is designed not only to respond to the questions posed by New York State, but also to make explicit the more general lessons that we hope will assist others.

The committee held a series of open meetings to obtain information from members of the public, representatives of affected counties, New York State agency staff, and the Siting Commission members and staff. For many people, discussing the siting process brought back painful memories and emotions. The Siting Commission had worked diligently at a process that it had undertaken in good faith to reach a result—siting a low-level waste facility—that it believed was important for the welfare of New York State. Citizens and officials of the affected areas, in turn, felt they had been forced to defend the interests and future health and well-being of their communities against a misguided, state-imposed process. Concerned about the very different interpretations of the process and the strong opinions held by various participants, we worked very hard to be objective about what we heard and read, and to be fair in what we wrote.

The committee appreciates the help it received from many New York State agencies and citizens. The New York State Department of Health, the agency sponsoring the study, and Steve Gavitt, our liaison with the department, provided supportive guidance and assistance to the committee throughout our review. Siting Commission members and staff cooperated willingly in providing information. They tried to understand our questions and answer them as accurately as possible given the passage of time since the work being reviewed had been carried out. Department of

Environmental Conservation staff were also helpful. Citizens and officials of Cortland and Allegany counties provided valuable comments to the committee that helped us better understand what had occurred from their perspective. The committee also thanks the New York State Geological Survey, in particular, Robert Fakundiny, state geologist, who participated in several of our meetings.

The committee is grateful for the aid of able and dedicated National Research Council staff. Ina Alterman helped during the formation of the committee and assisted us during the initial phases of our work. Charles Meade performed the herculean task of taking many pages of material written by a committee of 18 and turning it into a coherent, nonrepetitive draft. Kevin Crowley then refined the text and helped the committee clearly articulate its conclusions. Rebecca Burka kept the work well organized and the committee well cared for. Liz Landrigan and Scott Hassell assisted in the earlier stages of our project. Erika Williams assembled the many tables, figures, appendixes, and references; went through the report in great detail to check the numerous facts and figures; and assisted the committee with the preparation of the executive summary. We also appreciate the help of Cindy Monaco, former coordinator of the Cortland County Low-Level Radioactive Waste Office, who served as consultant to the committee. In this capacity she helped us identify additional information from affected counties and citizens and, through her experience in New York State, was able to provide some of the background history of citizens' involvement in the process.

Finally, I am personally very grateful for a fine committee whose members worked diligently and cheerfully to reach agreement in spite of viewing issues from a diverse set of disciplines and a wide range of experiences. It was a pleasure to serve as ringmaster for such a capable group.

Susan Wiltshire, *Chair*

Contents

Appendixes

Executive Summary

The 1986 New York State Low-Level Radioactive Waste (LLRW) Disposal Act set into motion a multiyear effort to identify a site for LLRW disposal in the state.[1] As required by this act, the governor of New York appointed a commission charged with siting low-level waste disposal facilities. The Siting Commission embarked on an ambitious effort to identify a site by performing a comprehensive screening of the land area of the state using a set of 60 technical and socioeconomic criteria.

By June 1990 the Siting Commission's work had come under intense public scrutiny and criticism from affected communities in the state. As a result, the 1986 act was amended to require a review of the Siting Commission's work by an independent panel of experts. The National Research Council (NRC) was asked to undertake this review, and a committee was appointed by the chairman of the NRC to review the Siting Commission's siting methodology. The committee's review is provided in this report.

REVIEW OF THE SITING PROCESS

The Siting Commission adopted a two-phase siting process, described in its *Plan for Selecting Sites for Disposal of LLRW*, hereafter referred to as the "Siting Plan." The first phase was designed to identify a small number of candidate sites by using existing data and limited reconnaissance studies. The second phase was designed to use on-site investigations to identify one or more sites for certification and licensing.

As charged in its statement of task (Appendix B), the committee addressed its review to the three steps in the first phase of the siting process that were completed by the Siting Commission before its activities were terminated by the governor of New York (Figure 1): (1) the exclusion of lands from consideration; (2) the selection of 10 candidate areas; and (3)

[1]Box 1.1 in Chapter 1 provides a brief discussion of LLRW in New York State.

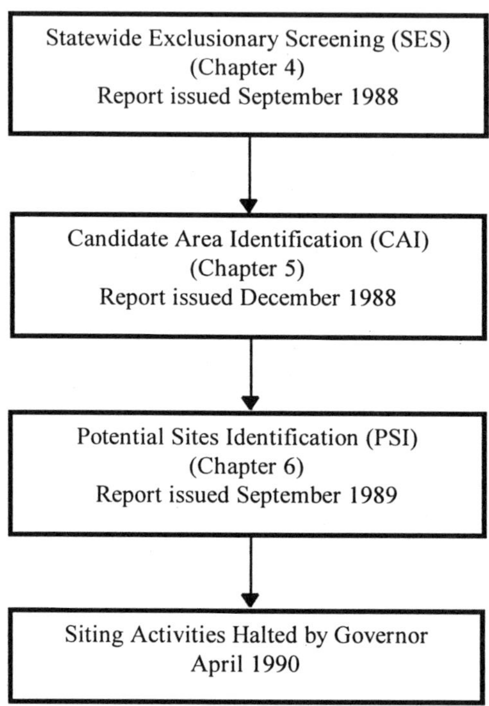

FIGURE 1 Major steps of the New York State siting process addressed in this report.

the selection of 5 potential sites. The committee also was directed to address eight specific questions regarding the technical and scientific adequacy of the siting criteria and site selection process. These questions are discussed in Chapter 8 of this report.

The siting process was guided by regulations issued by the New York State Department of Environmental Conservation (DEC). To address these regulations, the Siting Commission designed a set of 17 exclusionary and 43 preference criteria. Subsets of these criteria were used at each step of the siting process to screen the land area of the state. Exclusionary criteria were addressed to the nondiscretionary elements of law or regulation. Areas that did not satisfy these criteria were eliminated from consideration. Preference criteria were addressed to the discretionary elements of law or regulation. Subsets of these criteria were used to score and rank areas numerically during various screening steps. The scoring procedure was intended to guide the selection process toward areas and sites with desirable characteristics for an LLRW disposal facility. The Siting Commission used a Geographic Information System (GIS) to apply the criteria and calculate scores.

Statewide Exclusionary Screening

The first step in the screening process, Statewide Exclusionary Screening (SES), utilized a set of five exclusionary criteria to screen out areas of the state that were precluded by federal or state regulations from hosting an LLRW facility. The criteria applied at this step required no interpretation of regulations and employed data available for the entire state. This screening step excluded approximately 30 percent of the land area of the state.

In the committee's judgment, the exclusions in this step were applied appropriately and the criteria used were based on sound regulatory considerations. The Siting Commission had essentially no discretion in this step of the screening process; state law or regulation precluded a disposal facility from all of the excluded areas.

Candidate Area Identification

The objective of Candidate Area Identification (CAI) was to apply additional sets of exclusionary and preference criteria to the nonexcluded areas of the state in order to select a small number of areas for more detailed screening. Three sets of exclusionary and preference criteria were applied in this screening step:

1. Nine exclusionary criteria were applied to the nonexcluded lands from SES, resulting in the elimination of an additional 9.2 percent of the land area of the state.

2. The remaining lands were then scored and ranked using a set of 14 preference criteria, and 30 areas that exceeded a minimum cutoff score (4,400 out of a possible 5,000 points) were carried forward. A limited sensitivity analysis was conducted to evaluate the effect of certain weighting factors on scoring.

3. A qualitative screening was performed using a set of 5 exclusionary and 13 preference criteria to select 10 candidate areas.

The committee identified problems with the design or application of some of the exclusionary and preference criteria used in CAI screening, as noted in Chapter 5. These criteria address seismic hazards, existing mines, protection of ground water resources, buffers from water resources, best usage of surface waters, and air quality nonattainment.

The committee also identified several problems with the CAI screening process itself. The committee observed that the selection of candidate areas depended to a great extent on the sequence of screening steps and the combination of criteria applied at each step, rather than on the likely performance of the area as an LLRW disposal facility. The committee also concluded that the sensitivity analysis performed by the Siting Commission was poorly documented and had a number of technical shortcomings. In performing its own sensitivity analysis, the committee observed that an area must score favorably on a large number of criteria in order to exceed the cutoff score, which is a positive attribute of the screening process. The committee also observed, however, that the results of screening were biased toward rural areas and—a problem of particular concern—regions that lacked data. The committee noted that there is not a strong correlation between a score and the likely performance of an area as an LLRW disposal facility and, therefore, that the Siting Commission's use of cutoff scores in the CAI process was inappropriate.

Potential Sites Identification

The objective of the Potential Sites Identification (PSI) step of the screening process was to apply additional sets of exclusionary and preference criteria in order to identify five potential sites. PSI screening consisted of four discrete activities:

1. The candidate areas were screened using 13 exclusionary and 27 preference criteria to identify 96 sites.

2. The 96 sites were then screened using 1 exclusionary and 5 preference criteria. A total of 51 sites were retained.

3. Limited site inspections, performed by Siting Commission staff from inside their vehicles, were used to select 19 sites for further consideration.

4. The 19 sites were rescreened using the full set of exclusionary and preference criteria, and 5 potential sites were identified for further study.

During PSI screening, the Siting Commission also considered five parcels of land that had been offered by landowners as possible sites for an LLRW disposal facility, and included one of the parcels on the list of five potential sites.

The committee identified problems with either the design or the application of a few of the exclusionary and preference criteria used in the PSI screening step (Chapter 6). These criteria related to geologic complexity, subsurface dissolution, erosion, mineral soil groups, and subsurface drainage. The committee also identified problems with the PSI screening process itself. The first screening activity employed a cutoff score, even though site scores may be only weakly correlated with the likely performance of a site. The second screening activity was not part of the original Siting Plan, and it treated preference criteria as exclusionary conditions. With respect to the third screening activity, the committee found that the plan of the field surveys, the training of the staff performing the surveys, and the factors to be evaluated appear to have been conceptually sound. However, the committee was not provided enough information to assess whether the Siting Commission used the survey data appropriately. In one case the committee reviewed documentation suggesting that a site near the town of Taylor (the Taylor North site) should have been excluded based on survey data. In the fourth screening activity, selection of the five potential sites was based on a poorly documented staff decision. Performance and socioeconomic criteria were combined inappropriately in this screening step as well.

The committee also considered the Siting Commission's treatment of offered sites, especially the selection of the Taylor North site as one of the five potential sites. In the committee's judgment, the decision to include the Taylor North site among the five potential sites was inconsistent with the Siting Commission's policy that an offered site must be at least as good as other sites. The committee believes that the Taylor North site should have been excluded because of the presence of mineral soils in groups 1 through 4 under active agriculture and incompatible structures.

Although limited sensitivity analyses were conducted by the Siting Commission during PSI screening, they tested the influence of exclusionary criteria, which the Siting Commission had little freedom to change. In the committee's judgment, the Siting Commission should have performed a detailed sensitivity analysis to test the effects of preference criteria weighting and scaling factors on scoring. The committee performed its own sensitivity analyses and concluded that at least two of the steps had a bias toward rural areas. The analyses also suggested that there may be a weak correlation between a site's score and its suitability for LLRW disposal.

DISCUSSION

The failure of the New York siting effort can be attributed to both external and internal causes. The primary external cause was the unrealistic schedules imposed on the Siting Commission by the federal and state LLRW management acts. In its strict adherence to schedules, the Siting Commission failed to undertake the kind of careful planning and public outreach required to bring its work to a successful conclusion. The difficulty of the Siting Commission's task, however, was increased by the intensity of public opposition. Indeed, some opponents clearly and vocally stated that they would not accept any site, regardless of the technical justification, and some of them refused to participate in the siting process.

One of the internal causes of failure was ambiguity over the goal of the siting process. The Siting Plan noted that the goal of the screening process was "to identify sites that can satisfy the regulatory requirements and to demonstrate that no obviously superior alternatives can be readily identified" (Siting Plan, p. S-4). There may be many sites in New York that would meet the regulatory requirements to be certified and licensed. However, to demonstrate that a site is as suitable as all other readily identifiable alternative sites in the state, as implied by the language in the Siting Plan, is difficult at best. Many factors are involved in the suitability assessment, and no one site can be expected to be superior in all respects.

Part of this ambiguity may be due in part to perceptual differences among involved parties in what constitutes such a site. To the professionally trained members of the Siting Commission, staff, and contractor, an acceptable site could reasonably be perceived as one that meets the technical requirements for licensing and that is located in a community that will accept an LLRW disposal facility. Many of the affected communities, however, appeared to perceive such a site as "best" in some objective sense, and this perception may have been reinforced by the use of a highly technical screening process. The screening process, however, was sufficiently complicated that it would have been very difficult for the Siting Commission to demonstrate that sites superior to the ones chosen did not, in fact, exist.

The committee also believes that the technical deficiencies in the screening process contributed significantly to the failure of the siting process and that the sensitivity analyses performed by the Siting Commission were inadequate for identifying these problems. In addition, the committee also notes that the Siting Commission did not provide adequate documentation of some of its important screening decisions.

Another internal cause of failure was the lack of a strategic plan and quality assurance program. A good quality assurance program would have helped the Siting Commission identify critical data to validate its decisions and critical aspects of the siting process. It also would have provided a feedback mechanism to allow corrections to the siting program in a timely manner.

Finally, part of the failure of the siting process can be attributed to insufficient public participation. The early implementation of the Siting Commission's public participation plan was successful in that the public was informed, given an opportunity for input before decisions were made, and encouraged to comment on draft plans. As might be expected on the basis of other siting efforts, however, relatively few members of the public availed themselves of the opportunity to provide input at this stage of the process, probably because they did not connect this activity to their own communities.

Problems with public participation can be traced to the transition from the exclusion of lands to the selection of potential sites, beginning with CAI screening. The committee finds that the Siting Commission did not provide the leadership necessary to handle this transition well, probably because it lacked a strategic approach to site selection. The Siting Commission failed to recognize the change from an exclusionary to an inclusive role. That is, the Siting Commission did not redefine their role to one of finding a licensable site from one of excluding unlicensable areas. The committee also finds that the Siting Commission made critical siting decisions without appropriate public participation. Because the site selection methodology included many subjective elements, success or failure was very sensitive to the degree of public participation in Siting Commission activities.

LESSONS TO BE LEARNED

There are three important lessons to be learned from New York State's siting efforts. The first relates to the use of a "top-down" screening process that entailed a stepwise screening of the entire state to identify ever-smaller parcels of land from which a preferred site could be selected. In the committee's judgment, top-down screening should not be pushed beyond the capabilities of the data and selection criteria to support comprehensive and technically credible decisions. In the New York siting effort, top-down screening probably should not have been pushed much beyond SES, the only step for which statewide data of reasonably good quality were available and in which exclusionary criteria were based on laws and regulations viewed as reasonable by most parties. Attempting to screen out areas based on partial data sets leads to perceptions of unfair treatment, as areas with incomplete descriptive information (often rural regions) are retained for further review, and more thoroughly-described (often urban) areas are excluded, generally to the relief of those living there. Further, the need to create complex data sets to resolve such discrepancies can force the entire siting process toward expensive, time-consuming, and generally unattainable goals. Once data and criteria no longer support screening decisions, other analytical strategies should be considered, such as the collection of additional data or the adoption of a "volunteer" process.

The second lesson to be learned is that public acceptance of the process and the results is key to the success of a siting effort. Research has shown that nuclear waste evokes feelings of anxiety for many members of the public and results in special socioeconomic impacts in affected communities. To be successful, waste disposal siting efforts must be structured to address these effects through a high level of public involvement in facility siting, design, operation, and monitoring. This process must be a cooperative effort and requires the constructive participation of the potentially affected communities.

The third lesson to be learned is that strategic planning is essential to completing complex projects under tight deadlines. Such planning must include the identification of key milestones and objectives as well as procedures for dealing with the many unanticipated problems that inevitably arise in any project of this complexity. The strategic planning process could have led the commission to consider the need to change its approach at the turning point from exclusion of lands to selection of sites. Strategic planning could have impelled the Siting Commission to conceptualize the characteristics of an ideal site and also would have forced the Siting Commission to seek out and learn from previous siting efforts.

1
Introduction

In 1995 the New York State Legislature voted to end the activities of the New York State Low-Level Radioactive Waste Siting Commission (referred to throughout this report as the "Siting Commission," or "commission"). The Siting Commission had been in existence for eight years, and it had the primary responsibility for identifying a site and technology for disposal of low-level radioactive waste (LLRW) within New York State (see Box 1.1). Over the course of its history, the Siting Commission evolved from an obscure public body that held sparsely attended meetings to the focal point of fervent public protests that involved hundreds of citizens. Through September 1995, New York had expended about $55.2 million on regulation development and siting activities (Low-Level Radioactive Waste Forum, 1996).

Between 1988 and 1989, the Siting Commission attempted to carry out its mandate to identify licensable LLRW disposal sites. The process was set forth in state and federal laws; prescribed by state and federal regulations; and engaged a broad range of agencies, local governments, consultants, and private citizens. The Siting Commission was directed to consider the impacts on public health and safety; the nature of probable impacts on the environment, local economies, and governments; the adequacy of transportation routes; population densities surrounding the sites; and the ability to recover wastes at a later date.

The Siting Commission developed a statewide screening process based on an analysis of land use patterns, surface and ground water hydrology, geologic properties, demographic issues, and socioeconomic concerns. The objective was to exclude those regions precluded by law or regulations and then use a stepwise technical screening process to identify sites for detailed characterization. In the first of these tasks, the Siting Commission worked with little or no public interest. As the commission's work shifted from exclusion to site identification, however, its goals and methods became more controversial. The exclusionary phase was completed by September 1988 and resulted in the elimination of about 30 percent of the state from further consideration. By September 1989, the Siting Commission had identified five potential disposal sites in two different counties. The area of these sites totaled 3,896 acres, or approximately 0.01 percent of the land area of New York State.

In early 1990 the governor of New York suspended the Siting Commission's activities. At that time, intense public protests had disrupted the commission's work and prevented it from carrying out detailed evaluations of the five potential disposal sites it had identified previously. This aspect of New York State's experience was comparable to experiences throughout the nation. Because of widespread public concern regarding radioactive waste disposal, no new LLRW disposal sites have been opened in the United States, even though federal law required all regional compacts and noncompact states to have developed such capabilities by January 1, 1993 (see Chapter 2).[1]

In response to public pressure, the New York State Legislature amended the act that created the Siting Commission and the framework for identifying and licensing disposal sites. The new legislation mandated broad changes to the structure and procedures of the Siting Commission and significantly enhanced the opportunities for public participation in the siting process. The amended act also called for the creation of an independent technical and scientific panel to review the Siting Commission's work. Specifically, the amended act directed that

> . . . the department of health shall arrange to have one or more independent panels of technical and scientific experts review and evaluate the commission's decisions and report on its selection of a tentative preferred disposal method and decisions and report on lands excluded from consideration for siting permanent disposal facilities. . . . (L. 1990, C. 913, §3, amending New York State *Environmental Conservation Law* [ECL] Art. 29, §29-0303.11)

---

[1]The Ward Valley site in California has received a license for low-level waste disposal, but construction of the facility awaits the transfer of land from the U.S. Department of the Interior to the State of California.

BOX 1.1

New York State's LLRW—What is it and how much is there?

A good general description of low-level radioactive waste is given in the Nuclear Waste Primer (League of Women Voters, 1993), which notes that LLRW is

> a catchall category defined by what it is not rather than by what it is. [LLRW] includes all radioactive waste other than uranium mill tailings, transuranic waste, and high-level waste, including spent nuclear fuel. While most low-level waste is relatively short-lived and has low levels of radioactivity, some presents a greater radiation hazard. . . . The Nuclear Regulatory Commission classifies low-level waste into four groups according to the degree of hazard it poses and, consequently, the type of management and disposal it requires. Low-level waste that can be disposed of by shallow land burial is classified as A, B, or C, from least to most hazardous. States are responsible for the disposal of class A, B, and C waste.

See Appendix G of the present report for more detailed definitions of the classes of LLRW.

As projected in the Siting Commission's *1994 Source Term Report Executive Summary*, the LLRW disposal facility to be built in New York must isolate an estimated 4.3 million cubic feet (approximately 120,000 cubic meters) of LLRW—enough LLRW to fill a football field to a depth of about 75 feet (approximately 23 meters).

By volume, the state's LLRW is projected to consist of 98% Class A waste, 1% Class B waste, and 1% Class C waste. By activity (the rate at which a radioactive material emits radiation), the composition of the wastes is expected to be 8% Class A, 18% Class B, and 74% Class C.

According to the estimate, about 97% of the volume of waste generated within the state will be produced by nuclear power plants, with the remainder of the waste coming from medical and research facilities. The wastes come in a variety of forms ranging from paper, gloves, boots, resins, plastics, and metals, to ash from incinerated biological wastes and parts of decommissioned nuclear reactors. LLRW generators in New

York State are currently shipping their waste to the Barnwell disposal facility in South Carolina.

The facility to be built in New York is projected to receive LLRW for 60 years. Figure 1.1 shows how the amount of radioactivity contained in the disposal facility is expected to change over the life of the facility and beyond. For the purposes of the analysis, the facility was assumed to begin receiving LLRW in 1994. The peak in activity would occur about 36 years into the facility's life, when most of the decommissioning of nuclear power plants would be complete. After 100 years, less than 20% of the peak activity would remain.

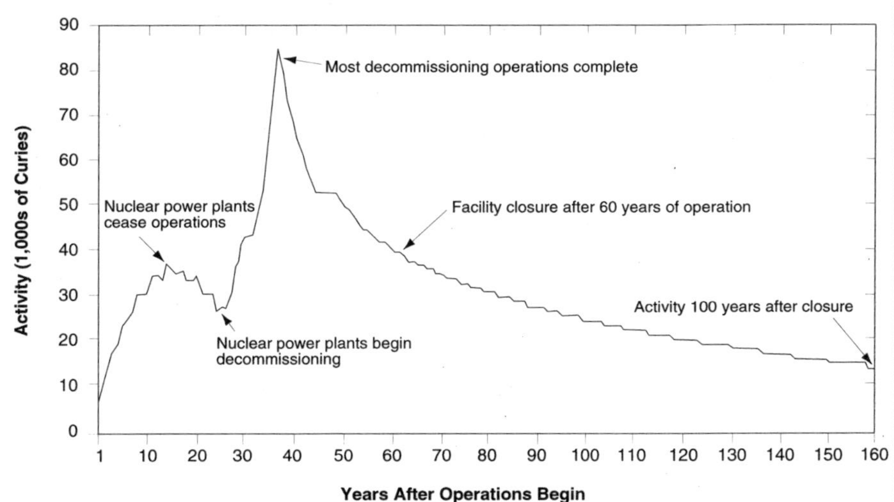

FIGURE 1.1 Expected activity through time of the LLRW to be accumulated in a disposal facility in New York State, assuming a 1994 opening date. Source: Siting Commission (1995).

In accordance with the amended act, the New York State Department of Health requested that the National Research Council (NRC) undertake a technical review of the siting process. The statement of task to the committee, which is included in Appendix B of this report, focused on eight questions that address the technical validity of the Siting Commission's screening process. The intent of the questions (and hence the committee's work) was to determine if the siting process and the decisions made by the Siting Commission were based on sound technical considerations and were consistent with good scientific practice.

The chair of the National Research Council appointed a committee of 21 experts to perform the technical review. This committee, which operated under the auspices of the NRC's Board on Radioactive Waste Management, included experts in many of the disciplines relevant to siting radioactive and hazardous waste facilities, including geology, hydrology, law, sociology, environmental health and science, radiation protection, soil science, mining engineering, and materials science. Biographical sketches of the committee members are given in Appendix C.

WORK PLAN

The committee met nine times between January 1994 and October 1995 to gather and review information, discuss the issues, and develop its report. Four of the meetings were held in New York, and parts of these meetings were open to the public and were advertised as such. The purpose of these open sessions was to obtain information relevant to the committee's review from Siting Commission members and staff, New York State agency staff, and interested members of the public. A list of presenters at the open sessions is given in Appendix D.

The committee also obtained information from several sources outside these meetings. A subgroup of the committee visited the offices of the Siting Commission during its November 1994 meeting to gather information on the Geographical Information System (GIS) procedures used in the siting process. In addition, the committee received and reviewed a large number of reports and memoranda from the Siting Commission, local governments, and citizens of New York. These documents are listed in Appendix E. The committee also received and reviewed written responses to a set of questions it submitted to the Siting Commission and the Department of Environmental Conservation. These questions are given in Appendix F.

Executive sessions of the committee were held to conduct the following business: (1) the conflict-of-interest discussion, which is required by the NRC of all committees; (2) discussion of the statement of task; (3) discussion of information and technical documents provided to the committee; (4) development of findings and recommendations; and (5) drafting of this report. Following established practices of the Board on Radioactive Waste Management, the parent board to this committee, these sessions were closed to all but NRC committee and staff members. Executive sessions were held at each of the four New York meetings. Five additional executive sessions were held at NRC facilities in Washington, D.C.; Irvine, California; and Woods Hole, Massachusetts.

SCOPE AND ORGANIZATION OF THIS REPORT

This report presents the committee's findings and conclusions concerning the Siting Commission's work through September of 1989 (the issue date of the *Report on Potential Sites Identification*) and addresses the first three parts of the statement of task (see Appendix B): (1) the exclusion of lands as potential sites, (2) the selection of 10 candidate areas, and (3) the selection of 5 potential sites. The statement of task directed the committee to address each of its tasks in a separate report. After extensive discussion, however, the committee decided to address these tasks in a single report to improve the presentation of findings and reduce redundancies.

The remaining two parts of the statement of task, a review of the disposal methodology and source term, were cancelled by the New York State Department of Health after the siting process was halted in 1995. Consequently, these tasks are not addressed in this report.

This report is organized to address explicitly the first three parts of the committee's statement of task. Chapter 2 provides the background to the work of the Siting Commission, including the federal and state legislative framework that guided the siting effort. Chapter 3 provides a general review and discussion of the Siting Commission's plan and methodology for selecting a disposal site. The Statewide Exclusionary Screening (SES) step of the siting process—the subject of the first part of the task statement—is reviewed in Chapter 4. Chapter 5 addresses the second part of the statement of task, Candidate Area Identification (CAI), and Chapter 6 addresses the third part of the statement of task, Potential Sites Identification (PSI). Concluding observations and the committee's recommendations about the siting process are presented in Chapter 7. Although the eight questions laid out in the charge to the committee are addressed throughout the report, as appropriate, the committee's responses to the questions are summarized in Chapter 8 for ease of reference.

2

Federal and State Low-Level Radioactive Waste Acts

The efforts of New York State to site and develop a low-level radioactive waste (LLRW) disposal facility were required by federal laws enacted in 1980 and 1985. These laws established the responsibility of each state for low-level radioactive waste disposal and set strict schedules for compliance with associated milestones and penalties. The New York State Legislature enacted a state law in 1986 to comply with the federal laws. This law mandated an aggressive effort to site and develop an LLRW disposal facility in the state. This chapter provides a brief review of these federal and state acts to set the context for New York State's siting efforts. Table 2.1 illustrates the relation in time between the legislative mandates and the major activities of the Siting Commission.

FEDERAL LLRW POLICY ACT OF 1980

The federal Low-Level Radioactive Waste Policy Act of 1980 (P.L. 95-573; hereafter referred to as the "1980 Act") required states to provide disposal capacity for commercial low-level radioactive waste[1] generated within their borders. (See Appendix H for a copy of the act.) Facilities for disposing of the nation's civilian low-level radioactive

---

[1]The 1980 Act defines low-level radioactive waste as "radioactive waste not classified as high-level radioactive waste, transuranic waste, spent nuclear fuel, or byproduct material as defined in section 11e.(2) of the Atomic Energy Act of 1954." For a classification of low-level radioactive waste according to Title 10, Part 61 of the *Code of Federal Regulations* (10 CFR 61), Licensing Requirements for Land Disposal of Radioactive Waste, see Appendix G. See also Box 1.1.

TABLE 2.1 Timeline Showing History of New York State LLRW Siting Process

1980	
Dec.	• 1980 Act takes effect
1986	
Jan.	• 1985 Amendments Act takes effect
July	• Federal milestone requiring compact legislation or state decision to develop facility
	• 1986 State Act takes effect
1987	
May	• Governor appoints Siting Commission
June	• First Siting Commission meeting
Dec.	• DEC issues draft 6 NYCRR 382 siting criteria regulations
	• Siting Commission issues draft Siting Plan
1988	
Jan.	• Federal milestone for siting plan and designation of host state
Mar.	• Governor appoints chairman of Advisory Committee
Sept.	• SESR issued
Nov.	• Siting Plan issued
Dec.	• CAIR issued
1989	
Sept.	• ROPSI issued
1990	
Jan.	• Federal milestone requiring filing of facility license applications or governor's certification of state's ability to manage waste
April	• Governor halts siting effort
July	• Amendments to 1986 State Act take effect, with requirement for independent review
1992	
Jan.	• Federal milestone requiring facility license applications
1993	
Jan.	• Federal milestone allowing compacts to limit facility access
Mar.	• DEC issues final 6 NYCRR 382
Apr.	• NRC study begins
Aug.	• *Excluded Areas Report* issued
1995	
Aug.	• New York State Legislature ends work of Siting Commission

NOTE: CAIR = *Candidate Area Identification Report*; DEC = New York State Department of Environmental Conservation; NRC = National Research Council; NYCRR = *New York Code of Rules and Regulations*; ROPSI = *Report on Potential Sites Identification*; SESR = *Statewide Exclusionary Screening Report*; Siting Plan = *Plan for Selecting Sites for Disposal of Low-Level Radioactive Wastes.*

waste previously had been licensed by the U.S. Nuclear Regulatory Commission (USNRC) and agreement states[2] and operated by commercial firms.[3] In the late 1970s the states hosting these facilities became concerned about corrosion and leakage of waste packages and expressed the need for geographic equity in the disposal of low-level waste.

The 1980 Act encouraged states to form regional compacts, subject to approval by the U.S. Congress, and to share disposal facilities. As an incentive to form compacts and to expiate potential interstate commerce clause issues, the 1980 Act allowed compacts to exclude wastes generated outside their borders (i.e., "out-of-region wastes") after January 1, 1986.

The National Governors' Association, the National Conference of State Legislatures, and many state governments supported the 1980 Act, anticipating an opportunity to demonstrate the ability of the states to solve problems relating to the safety and welfare of their citizens and their economies. After enactment of the 1980 Act, many states entered into serious negotiations to form regional compacts. The states with operating LLRW disposal facilities—Nevada, South Carolina, and Washington—were quickly joined by nearby states to form the Rocky Mountain Low-Level Radioactive Waste Compact, Southeast Interstate Low-Level Radioactive Waste Management Compact, and Northwest Interstate Compact on Low-Level Radioactive Waste Management, respectively. Some states, including several populous states with large volumes of low-level waste, opted not to join a compact but to develop their own facilities.

---

[2]See discussion under regulatory requirements later in this chapter.

[3]Prior to the 1980 Act, commercial sites were in operation at Beatty, Nevada; Barnwell, South Carolina; and Richland, Washington. Disposal sites had been operated at Maxey Flats, Kentucky; Sheffield, Illinois; and West Valley, New York, but they were forced to close because of operational problems. The West Valley site was closed after the disposal trenches filled with water. The New York State Department of Environmental Conservation regulations now prohibit waste disposal in unlined trenches (Title 6, Part 382 of the *New York Code of Rules and Regulations*, Regulation of Low-level Radioactive Waste Disposal Facilities: Certification of Proposed Sites and Disposal Methods, Section 382.31(a)(1)).

New York entered into negotiations with other northeastern states under the sponsorship of the Coalition of Northeastern Governors[4] in an effort to establish a northeast regional compact. Because these states produced disparate volumes of low-level radioactive waste, and because of other political and technical factors, this 11-state regional compact was never established. After a thorough study, concluded in April 1984, the New York State Energy Office recommended that the state not join a regional compact but instead enact legislation, with adequate appropriations, to establish a two-year process to identify a site for permanent LLRW disposal within New York. In 1985 only New Jersey and Connecticut joined the Northeast Interstate Low-Level Radioactive Waste Compact. Subsequently in February 1986, Pennsylvania established the Appalachian States Low-Level Radioactive Waste Compact, now composed of Delaware, Maryland, Pennsylvania, and West Virginia. Pennsylvania agreed to be the host state.

FEDERAL LLRW AMENDMENTS ACT OF 1985

In the years immediately following passage of the 1980 Act, states made relatively little progress in developing new LLRW disposal facilities. By early 1985 it was clear that no new disposal capacity would be available by January 1, 1986—the date specified in the 1980 Act when compacts with existing facilities could exclude out-of-region wastes. In recognition of this problem, the U.S. Congress passed the Low-Level Radioactive Waste Policy Amendments Act of 1985 (P.L. 99-240; hereafter referred to as the "1985 Amendments Act," see Appendix H). To facilitate the passage of this act, the three states with existing disposal facilities agreed to remain open to the rest of the nation for an additional seven years (from 1986 to 1992). In return, these states were authorized to impose surcharges on waste received from generators outside their compacts. In addition, Congress established strict milestones for the remaining states to develop LLRW disposal facilities.

---

[4]The Coalition of Northeastern Governors included the states of Connecticut, Delaware, Maine, Maryland, Massachusetts, New Hampshire, New Jersey, New York, Pennsylvania, Rhode Island, and Vermont.

An escalating set of penalties was to be imposed for failure to meet the milestones of the 1985 Amendments Act. Based on a historical analysis, limits were placed on the volumes of waste that could be disposed of at existing facilities between 1986 and 1992. LLRW generators in states that failed to meet the milestones of the 1985 Amendments Act were subject to being assessed a penalty surcharge or denied access to existing disposal facilities. The milestones and associated penalties are described below.

- *July 1, 1986.* Each state was required to enact compact legislation, or the governor was required to indicate the state's intent to develop a disposal facility. Failure to meet this milestone could result in a penalty surcharge on generators of $20 per cubic foot of waste disposed. Continued failure to meet this milestone by January 1, 1987, could result in denial of access to existing disposal facilities.
- *January 1, 1988.* Each state was required to develop and delegate the authority to implement a siting plan. Failure to meet this milestone could result in a penalty surcharge on generators of $40 to $80 per cubic foot of waste disposed. Continued failure to meet this milestone by January 1, 1989, could result in denial of access to existing disposal facilities.
- *January 1, 1990.* Each state was required to file a complete license application or the governor was to certify that the state would provide for the disposal of its waste. Failure to meet this milestone could result in denial of access to the existing disposal facilities.
- *January 1, 1992.* Each state was required to file a complete license application with the appropriate state or federal authority. Failure to meet this milestone could result in a penalty surcharge on generators of $120 per cubic foot of waste disposed.
- *January 1, 1993.* Compacts with existing disposal sites could prohibit disposal of wastes from outside their regions.

Effective January 1, 1993, all compacts were authorized to exclude out-of-region waste from their disposal facilities. Further, if a state or compact failed to provide disposal capacity by that date, generators could require states to "take title to" the waste generated within their borders. If a state or compact failed to develop a new disposal facility by January 1, 1996, the state was obligated to take title to and possession of its waste. This provision was later struck down by the U.S. Supreme Court, as discussed below, while other provisions were upheld.

The 1985 Amendments Act also clarified the wastes for which states would be responsible. These included all LLRW generated within the state that consisted of or contained class A, B, or C radioactive waste as defined by 10 CFR 61.55,[5] in effect on January 26, 1993, except for waste owned or generated by the U.S. Department of Energy, waste owned or generated by the U.S. Navy as a result of the decommissioning of U.S. Navy vessels, or waste from the Formerly Utilized Sites Remedial Action Program.

The State of New York, joined by Allegany and Cortland counties, challenged the constitutionality of the 1985 Amendments Act before the U.S. Supreme Court in 1992. The court upheld the constitutionality of the act and the authority of the compacts to prohibit disposal of out-of-region wastes; however, it struck down the provision that required states to take title to and possession of their wastes. The court reasoned that this provision exceeded the powers of Congress and that it was inconsistent with the Tenth Amendment to the Constitution of the United States.[6]

Implementation of the 1985 Amendments Act

From 1986 to late 1992, the three states and compacts with existing disposal facilities worked together to enforce the surcharge and access penalties mandated by the 1985 Amendments Act. These states and compacts did in fact deny access to their disposal facilities for failure to meet 1985 Amendments Act milestones. The Beatty, Nevada, site ceased operation altogether on December 31, 1992. The Northwest

[5]Title 10, Part 61, *Code of Federal Regulations*, Licensing Requirements for Land Disposal of Radioactive Waste. See Appendix G.

[6]The Tenth Amendment states, "The powers not delegated to the United States by the Constitution, nor prohibited by it to the States, are reserved to the States respectively, or to the people."

Compact limited waste acceptance at its Richland, Washington, facility to member states effective January 1, 1993. The Northwest Compact also entered into an agreement with the Rocky Mountain Compact to accept its waste. As of March 1996, 44 states were members of 10 compacts (Figure 2.1).

Effective July 1, 1994, the Southeast Compact also limited waste acceptance at its Barnwell, South Carolina, site to member states. In mid-1995, however, the South Carolina Legislature passed legislation to withdraw from the Southeast Compact and to make the Barnwell facility available, with the assessment of a very large surcharge, to all states in the nation except North Carolina. The legislation took effect on July 1, 1995. The Barnwell facility contains enough licensed space to remain open for 10 years; however, availability of the space depends upon uncertain political factors that could shorten significantly the life of the facility. Several other facilities are open to all states for treatment and/or disposal of specific types of wastes such as naturally occurring radioactive materials (NORM), soil and debris from cleanup of contaminated sites, liquid scintillation wastes, and certain other types of low-activity, low-level radioactive wastes.

NEW YORK STATE LLRW MANAGEMENT ACT OF 1986

In July 1986 the governor of New York signed the Low-Level Radioactive Waste Management Act (hereafter referred to as the "1986 State Act"), which established a framework for complying with the federal 1985 Amendments Act. The 1986 State Act established a Siting Commission and an Advisory Committee and assigned specific functions to several state agencies for establishing a disposal facility (see Appendix H). This legislation contained several explicit directives for the rapid establishment of a disposal facility, including the following:

> The legislature finds immediate implementation of steps toward establishing by January first, nineteen hundred

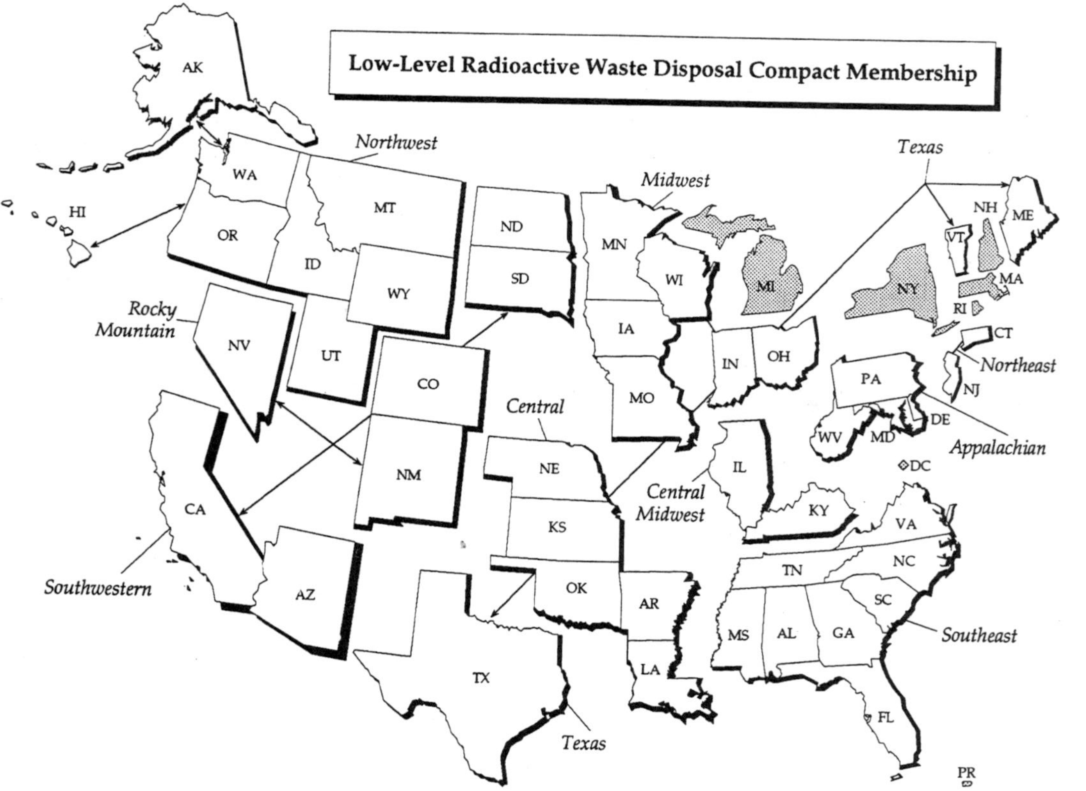

FIGURE 2.1 Map of the United States showing the status of low-level radioactive waste compacts as of March 1996. Shaded states do not belong to a compact. Source: Graphic by Afton associates for the *LLW Forum*. March 1996.

> ninety-three low-level radioactive waste management facilities . . . necessary to provide for continued operation of essential and beneficial . . . facilities in New York which use radioactive materials. (L. 1986, C. 673, § 2)

> The commission shall immediately commence the preparation of a siting and disposal method selection. . . . (L. 1986, C. 673, § 5, amending New York State *Environmental Conservation Law* [ECL] Art. 29, § 29-0303.1)

Agency Responsibilities

The New York Legislature split the responsibilities to carry out the mandates specified in the 1986 State Act among several state agencies. A list of responsible agencies and tasks is shown below:

• *New York State LLRW Siting Commission.* The Siting Commission was directed to select sites and disposal methods, to prepare a draft environmental impact statement, and to apply to the New York State Department of Environmental Conservation (DEC) for certification of site(s) and method(s). Members of the Siting Commission were to be appointed by the governor and were to include a geologist, physician, health physicist, professional engineer, and a private citizen who would act as chairperson.

• *New York State Energy Research and Development Authority.* The legislature directed the New York State Energy Research and Development Authority to design and operate the disposal facility; to participate in preparation of a draft environmental impact statement on facility design, construction, operation, and closure; to apply to DEC for permits to build, operate, and eventually close the facility; and to apply to the New York State Department of Labor for a license to handle radioactive materials. The New York State Energy Research and Development Authority was also required to collect information and provide annual reports to the governor and legislature on the types and quantities of LLRW generated within the state and to collect annual assessments from operating nuclear power plants.

• *New York State Department of Environmental Conservation.* The legislature directed the DEC to promulgate regulations specifying criteria for identification of a disposal site and disposal method; certify compliance with these regulations; prepare a final environmental impact statement on facility design, construction, operation, and closure; and approve or deny permits for construction and operation. The DEC was also required to promulgate regulations on financial assurance requirements for LLRW disposal facility operators and to make recommendations about the nature and form of state assistance to local communities. Upon reviewing the application from the Siting Commission, the DEC would certify the site if it was found to comply with DEC regulations.

• *New York State Department of Health.* The New York State Department of Health was required to conduct a statewide public information program on the health and safety implications of LLRW management.

• *New York State LLRW Advisory Committee* (referred to throughout this report as the "Advisory Committee"). The purpose of the Advisory Committee was to advise the Siting Commission on the selection of the site and disposal methods, to advise DEC on its LLRW regulatory program, and to advise the New York State Department of Health on its LLRW public information program. The legislature directed the governor to appoint the members and the chair of the Advisory Committee. The members included the state geologist; the commissioners of health, labor, the state Energy Office, and transportation, or their respective designees; the secretary of state, or his or her designee; two representatives of nonprofit environmental organizations; two health physicists or physicians; two representatives of LLRW generators; and one private citizen. Additionally, three private citizens were to be appointed to the Advisory Committee from each of the counties in which proposed sites were identified.

Site Development

The 1986 State Act did not specify the details of the site development process. It encouraged the selection of a single site, but a second site was allowed if the Siting Commission found

> that the use of an additional site presents specific advantages. . . . To the extent the commission determines that different disposal methods are appropriate for different categories of low-level radioactive waste with differing physical or chemical characteristics, the commission may select more than one disposal method to be utilized at each particular site. (L. 1986 C. 673 § 5, amending ECL Art. 29, § 29-0303.5)

The 1986 State Act reflected the strict schedule provisions of the federal 1985 Amendments Act by directing the Siting Commission to

> immediately commence the preparation of a siting and disposal method selection which shall, upon certification by the department [DEC], be the site or sites and method or methods for permanent disposal facilities. (L. 1986, C. 673, § 5, amending ECL Art. 29, § 29-0303.1)

Further, the Siting Commission was directed to

> complete the preparation of its site and disposal method selection and . . . submit its application for certification of this selection by the department [DEC] . . . by December 1, 1988. (L. 1986, C. 673, § 5, amending ECL Art. 29, § 29-0303.2)

From the effective date of the legislation, New York State had 28 months to select a low-level waste disposal site or sites and to prepare a draft environmental impact statement on each.

Public Participation

The 1986 State Act specifically called for public participation throughout the process of site selection. Opportunities for citizen participation were to begin with the development of the site selection plan and continue through each of the decision points leading to the selection of a preferred site.

The act distributed the responsibilities for public participation among the Siting Commission, New York State Department of Health, and the Advisory Committee. The Siting Commission was required to

> keep the public informed of its activities in developing the draft environmental impact statement . . . and encourage the public to participate by providing views, comments, information, and analysis concerning siting and disposal method selection for permanent disposal facilities. (L. 1986, C. 673, § 5, amending ECL Art. 29, § 29-0307)

The Advisory Committee was established as a specific form of public input. It was directed by the New York State Legislature to

> provide information and review and assist activities of the commission . . . and to receive a written report from the commission, the department [DEC], and the energy research and development authority on plans and progress in carrying out activities and duties pursuant to the low-level waste management act. In particular, the advisory committee shall provide information and recommendations in response to the commission's draft environmental impact statement . . . including reviewing public views, comments, and information submitted in response thereto. (L. 1986, C. 673, § 5, amending ECL Art. 29, § 29-0501.2.a)

The New York State Department of Health was also given responsibility for public participation in the following directive:

> In consultation and cooperation with the advisory committee . . . and with the department of environmental conservation, the commissioner of health shall plan and carry out a statewide public information program on the public health and safety implications of low-level radioactive waste management. The content of such program shall include a basic explanation of the types of materials which comprise low-level radioactive waste and why such materials require special handling and care; and reasonably detailed explanations of alternative disposal methods and their probable effects. (L. 1986, C. 673, § 6, amending ECL Art. 24-C, § 2485)

The 1986 State Act required DEC to hold public hearings on the proposed siting criteria regulations (L. 1986, C. 673, § 5, amending ECL Art. 29, § 29-0103.2) and on the draft environmental impact statement as well as the application for certification (L. 1986, C. 673, § 5, amending ECL Art. 29, § 29-0105.1). The final environmental impact statement was to include copies of the public hearing minutes; the recommendations of the Advisory Committee; and the department's responses to the views, comments, and recommendations.

Technical Guidelines

The 1986 State Act provided the Siting Commission with several technical guidelines for siting a low-level radioactive waste disposal facility. The relevant text from the act is given below:

> The commission shall take into account the following factors in the selection of the permanent disposal facility site or sites and disposal method or methods:

a) the nature and probability of the impacts on public health and safety, including predictable adverse effects from: (i.) accidents during transportation of low-level radioactive waste to such facilities; (ii.) contamination of ground or surface water by leaching and runoff from such facilities; and (iii.) fires or explosions from improper storage or disposal of volatile, combustible, or potentially explosive materials, if any, which may compose a portion of the low-level radioactive waste to be delivered to such facilities;

b) the nature of the probable environmental impacts, including the predictable adverse effects on the natural environment and ecology, scenic, historic, cultural and recreational values, water and air quality, and wildlife;

c) the potential for avoidance or mitigation of harm from the unanticipated release of low-level radioactive waste or contaminated materials;

d) the ability for retrieval or recovery of such waste;

e) differences in the density of population in the vicinity of the potential sites;

f) the adequacy of routes and means for transportation of low-level radioactive waste to such facilities;

g) the nature of the probable impact of such facilities on local governmental units within which such facilities would be located; and

h) the comparative economic implications, including those resulting from engineering considerations, of the potential site or sites and disposal methods for such facilities. (L. 1986, C. 673, § 5, amending ECL Art. 29, § 29-0301.4–4(h))

Implementation

Although the 1986 State Act took effect in July 1986, an active effort to site a disposal facility did not begin until about 10 months later—in May 1987—when the governor appointed the five-member Siting Commission (Table 2.1). The executive director of the Siting Commission was appointed by the chairman of the Siting Commission, and the professional staff of the Siting Commission were hired soon thereafter. Employees of the Siting Commission collectively represented the following professions: law, geology, hydrology, LLRW disposal technology, health physics, environmental analysis, communications, and quality assurance. The Siting Commission held its first meeting in June 1987. The governor appointed the chairman of the Advisory Committee in March 1988, about nine months after the first Siting Commission meeting.

REGULATORY REQUIREMENTS

In 1982 the USNRC promulgated a set of regulations to govern the land disposal of LLRW at all commercial facilities (10 CFR 61). The regulations set forth the procedures for obtaining a license for a land disposal facility, performance objectives to protect human health and the environment, technical requirements for facilities, and financial assurances. They describe the roles of states and Indian tribes, records, reports, tests, and inspections and address site characterization, disposal site suitability requirements, and disposal site design.

In 1959 the federal Atomic Energy Act authorized the USNRC (and its predecessors) to discontinue regulatory authority over radioactive materials in states that had complied with USNRC regulations and that had established adequate programs to protect public health and safety. States that have met these criteria are called “agreement states.” Because New York has been an agreement state since 1962, a low-level radioactive waste disposal facility in the state would be licensed by the New York State Department of Labor, the agreement state agency. The regulations for the licensing process were developed by the DEC and were consistent with those of the USNRC.

DEC Responsibilities

As required by the 1986 State Act, the DEC issued draft regulations with specific criteria to site a permanent disposal facility in New York in December 1987 (6 NYCRR Part 382[7]). The DEC also conducted an environmental impact statement process for the adoption of the Part 382 regulations, and the final environmental impact statement was issued with the proposed regulations. The environmental impact statement analyzed the impact of the regulations and numerous issues associated with various provisions in Part 382.

The DEC regulations "establish[ed] minimum criteria needed to ensure long-term isolation of LLRW and control exposure of the public and the environment to radiation from the disposal of LLRW" (DEC, 1987, p. 1-4). These regulations established detailed facility siting criteria, which are discussed further in Chapters 3, 4, 5, and 6 of this report. (They also established facility design criteria, which are not covered in this report for reasons discussed in Chapter 1.) The 1986 State Act also required the DEC to promulgate rules governing the financial assurance requirements for the disposal facility and regulating the design, construction, monitoring, and closing of the facility. The DEC regulations became a starting point for the Siting Commission's efforts to develop a plan for site identification and for selection of a disposal method. In order to be licensed, a proposed site was required to meet the DEC regulations.

Public Participation

The DEC regulations were much less specific than state law on the matter of public participation. They suggested, but did not require, that "[in] developing guidance [to the Siting Commission on developing plans for site characterization studies] the Department [DEC] may seek

---

[7]Title 6, Part 382, *New York Code of Rules and Regulations*, Regulation of Low-Level Radioactive Waste Facilities: Certification of Proposed Sites and Disposal Methods.

and consider recommendations from the public" (6 NYCRR 382.6(b)(2)). In addition, the regulations specifically required the Siting Commission to "hold at least one public scoping meeting to receive public comments on the scope of a draft environmental impact statement to be prepared regarding the Siting Commission's application for certification" (6 NYCRR 382.6(c)(2)).

Siting Criteria

The Part 382 regulations of the DEC are divided into seven subparts: (A) general provisions; (B) licenses; (C) performance objectives; (D) technical requirements for land disposal facilities; (E) financial assurances; (F) participation by state governments and Indian tribes; and (G) records, reports, tests, and inspections. The majority of the present report deals with Subpart D, which sets forth the requirements for siting of a land disposal facility.

ANALYSIS AND DISCUSSION

The 1986 State Act imposed a strict schedule on the Siting Commission and other state agencies in order to keep New York in compliance with the federal 1985 Amendments Act. The Siting Commission was directed to select a permanent disposal facility "after consideration of all relevant public health and safety, environmental and economic factors" (L. 1986, C. 673, § 5, amending ECL Art. 29, § 29-0303.5). The Siting Commission was required to complete the site and disposal method selection, along with a draft environmental impact statement, and submit them for certification no later than December 1, 1988 (L. 1986, C. 673, § 5, amending ECL Art. 29, § 29-0303.2)—that is, about 19 months after its members were appointed by the governor.

The New York State Legislature directed the Siting Commission to complete an extensive list of tasks during this 19-month period. These tasks included selecting potential sites, performing site characterization studies and monitoring, and submitting plans for the characterization studies to DEC for prior review. The Siting Commission was also required to consult with DEC regarding the form and content of the application for certification; to submit, for the DEC's approval, its proposed scope before preparing the draft environmental impact statement; to hold public meetings on the scope of the draft environmental impact statement; and to submit copies of the application to the DEC, the New York State Energy Research and Development Authority, the legislature, and the governor.

As the committee shows later in this report, the Siting Commission did not appear to question this ambitious schedule but, rather, made every effort to meet the milestones. Indeed, state and federal schedules appear to have driven many of the priorities set and decisions made by the Siting Commission. The focus on deadlines may have created pressure to move too quickly during various stages in the site selection process. Subsequent chapters show that the compressed schedules in combination with insufficient strategic planning were detrimental to the effectiveness and credibility of the Siting Commission's work and contributed to the failure of the siting process.

3

Overview of the Siting Process

This chapter provides a general description and analysis of the site selection process formulated by the Siting Commission to meet its responsibilities under the 1986 State Act, described in Chapter 2. New York State submitted a draft of its site selection plan to the U.S. Department of Energy to meet the January 1, 1988, milestone for the development of a siting plan set by the 1985 Amendments Act (see Table 2.1). The Siting Commission followed this plan until the process was halted by the governor in 1990.

The siting process involved several discrete screening steps. This chapter describes how each of the screening steps fit into the overall process and provides an analysis of those characteristics common to many of the individual steps. Subsequent chapters provide a more detailed analysis of several of the discrete steps.

SITING PLAN

The framework for the siting process is presented in the November 1988 Siting Commission's *Plan for Selecting Sites for Disposal of Low-Level Radioactive Wastes* (referred to throughout this report as the "Siting Plan"). The Siting Plan describes the goal of the siting process as follows (Siting Plan, p. 2-4):

> The process will identify sites that are expected to be potentially suitable for low-level radioactive waste disposal. It may not, however, identify every possible site within the state. Moreover, the process is intended to identify sites that can satisfy the site suitability criteria of 6 NYCRR Part 382 and to demonstrate that no obviously superior alternatives can be identified.

The Siting Plan acknowledges that other factors besides technical suitability are necessary for certification (p. 2-4):

> The Siting Commission recognizes that the selected site must be suitable for certification from the standpoints of technical suitability, public acceptance, and policy considerations. The decision-making process must balance the geologic, hydrologic, and environmental considerations with social, economic, and policy needs.

The Siting Commission makes it clear in the siting plan, however, that technical suitability is the primary focus of the screening process (Siting Plan, p. 2-1):

> The Siting Commission's objective is the selection of sites that are technically suitable for protecting public health and safety and the environment. Once technically suitable sites are identified, the Siting Commission will then be able to take into consideration other factors in selecting candidate sites.

The Siting Plan established a two-phase approach to selecting a low-level radioactive waste (LLRW) disposal site (Figure 3.1). Phase 1 was designed to identify a small number of potentially certifiable sites[1] based on a screening process that employed existing data or limited reconnaissance studies. Phase 2 called for detailed, on-site studies to identify a single "preferred site" for certification. This phase was never reached due to termination of the siting process by the governor of New York.

---

[1]The terms *certifiable* and *licensable* are used interchangeably throughout this report, although they have slightly different meanings. Sites that meet the New York State Department of Environmental Conservation (DEC) regulations for disposal of radioactive waste are referred to in this report as *certifiable sites*—although actual certification would be granted only after a review of an application by the DEC. Sites that meet the DEC regulations are also referred to as *licensable sites*—although, as noted in Chapter 2, the responsibility for issuing licenses resides with the New York State Department of Labor.

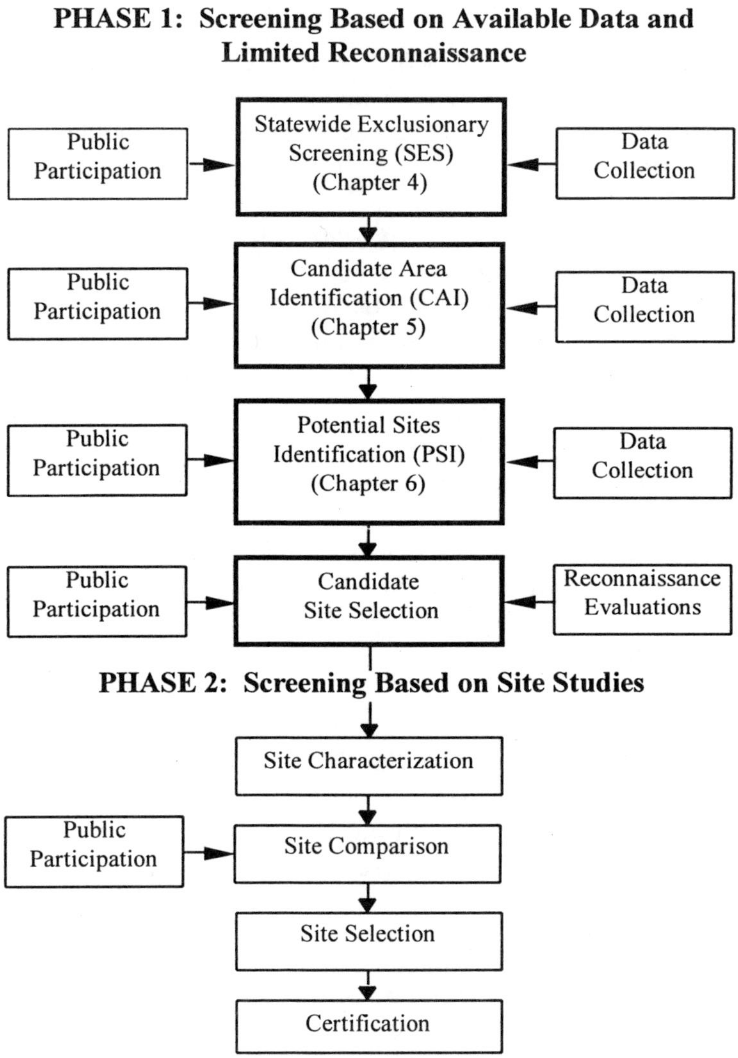

PHASE 1: Screening Based on Available Data and Limited Reconnaissance

Public Participation → Statewide Exclusionary Screening (SES) (Chapter 4) ← Data Collection

↓

Public Participation → Candidate Area Identification (CAI) (Chapter 5) ← Data Collection

↓

Public Participation → Potential Sites Identification (PSI) (Chapter 6) ← Data Collection

↓

Public Participation → Candidate Site Selection ← Reconnaissance Evaluations

PHASE 2: Screening Based on Site Studies

↓

Site Characterization

↓

Public Participation → Site Comparison

↓

Site Selection

↓

Certification

FIGURE 3.1 The New York State LLRW siting process.

The first phase of the screening process was subdivided into four parts:

1. *Statewide Exclusionary Screening* (SES), in which the Siting Commission screened the entire state using a set of *exclusionary criteria* to eliminate from consideration as an LLRW facility those areas prohibited from such use by law or regulation. This step is discussed in more detail in Chapter 4 of this report.

2. *Candidate Area Identification* (CAI), in which the nonexcluded areas were screened using a set of exclusionary and *preference criteria* to select about 10 "candidate areas" for more detailed study. This step is discussed in more detail in Chapter 5 of this report.

3. *Potential Sites Identification* (PSI), in which the candidate areas were screened using another set of exclusionary and preference criteria, to identify 4 to 8 "potential sites" for even more detailed study. This step is discussed in Chapter 6 of this report.

4. *Candidate Site Selection*, in which the Siting Commission was to use limited on-site studies to eliminate from consideration those potential sites having conditions that would preclude site certification and to select up to 4 "candidate sites" for detailed site characterizations. These detailed characterizations were to occur during phase 2 of the screening process.[2]

The Siting Plan noted that the DEC had the responsibility for developing the regulations and siting criteria governing the selection of a permanent disposal facility.[3] The Siting Plan indicated that the Siting Commission would carry out site selection in compliance with DEC's

[2]As noted previously, the siting process was halted by the governor of New York during the first phase of the screening process, after five potential sites had been identified.

[3]The regulations are laid out in Title 6, Part 382, *New York Code of Rules and Regulations* (NYCRR), the final version of which was issued in March 1993. A draft of these regulations and criteria was published 11 months prior to final version of the Siting Plan, in December 1987 (see Table 2.1). According to a January 21, 1994, memo to the committee from Mr. William Gilday, environmental analyst of the DEC Bureau of Radiation, these 1993 regulations "for the most part, were effective in March of 1988. These were substantially what the Siting Commission had to work with during the initial site selection activities."

regulations and associated criteria and that the Siting Commission "intend[ed] to communicate early and regularly with DEC to obtain guidance for ensuring that site selection activities comply with 6 NYCRR Part 382" (Siting Plan, p. 1-5).

As noted in Chapter 2, the commission had considerable discretion in selecting the methodology and the process to identify a certifiable site. Indeed, the Siting Commission noted that it needed guidance in interpreting some of the regulations. Notably, however, the DEC did not perceive that it should play an active role in guiding the site selection process. In response to this committee's questions regarding its role in the early stages of the commission's site selection plan (Appendix F), the DEC responded as follows:

> Because DEC would eventually review the Siting Plan and its implementation under that regulatory review, it was not appropriate for DEC to take a more active role in the SC's [Siting Commission's] implementation of the plan.

Screening Criteria

The DEC regulations in 6 NYCRR Part 382 identified a set of 13 factors that were important for siting an LLRW facility. The factors are shown in the first column of Table 3.1. The Siting Commission designated two sets of screening criteria to address these siting factors. The first set of criteria, *exclusionary criteria,* contained the mandatory elements (e.g., "the site must not be located . . .") of the regulations. Areas that failed to meet these criteria were disqualified from further consideration. The second set of criteria, *preference criteria,* included the discretionary elements (e.g., "sites are preferred that . . .") of the DEC regulations. The Siting Commission used these preference criteria to rank areas in terms of their favorability for an LLRW facility. As shown in Table 3.1, the Siting Commission designated a total of 17 exclusionary criteria and 43 preference criteria. A complete list of these criteria is given in Appendix I of this report.

TABLE 3.1 Preference and Exclusionary Criteria

Siting Factors	Number of Exclusionary Criteria	Number of Preference Criteria	Total Weight of Preference Criteria[a]
Geology	1	4	135
Natural resources	1	3	60
Ground water hydrology	1	3	135
Surface water hydrology	3	5	150
Meteorology and climatology	0	3	55
Air quality	1	1	30
Ecology	1	1	55
Incompatible nearby activities	0	2	40
Demographic patterns	1	3	115
Land use	6	5	70
Cultural resources and aesthetics	2	3	45
Transportation	0	7	75
Socioeconomics/ community services	0	3	35
Total	17	43	1,000

[a]Weights are discussed later in this chapter.

The use of the two different categories of screening criteria was initiated by the Siting Commission and was not required by the DEC regulations. In response to a question on this subject from the committee (Appendix F), the Siting Commission noted that this approach had been suggested by the senior staff of its primary contractor for the site selection process, Roy F. Weston, Inc.

Except for statutory exclusions, the decisions regarding classification of criteria as exclusionary or preference were made at the discretion of the Siting Commission. In an appendix the Siting Plan summarizes the statutory or regulatory basis for each of the criteria. Because the exclusionary and preference criteria played a central role in the site selection process, much of this report focuses on the design and application thereof.

Disposal Methodology

Although not required by the 1986 State Act, the Siting Plan indicated that the selection of a disposal method would occur concurrently with the selection of candidate sites. The Siting Plan stated (Siting Plan, p. 2-29):

> . . . up to three disposal methods and up to four candidate sites will be developed and evaluated for compatibility. Combinations that are incompatible will be eliminated from further consideration. The remaining site/disposal method combinations will be retained for further evaluation.

The Siting Plan indicated that there would be interaction between the selection of a site and the selection of a disposal method throughout the siting process.

In the Siting Plan the Siting Commission considered two different disposal technologies: (1) aboveground or belowground disposal and (2) mine disposal. The first method would involve the construction of engineered containment structures at, or slightly below, ground level. The second method would use existing or purpose-built

TABLE 3.2 Disposal Method-Specific Exclusionary and Preference Criteria (criterion numbers are shown in parentheses)

Type of Criterion	Aboveground/Belowground Disposal	Mine Disposal
Exclusionary	Wetlands exclusion (17)	Existing mine exclusion (4)
	Surface water bodies (15)[a]	
Preference	Unconsolidated stratigraphic units (14)	Existing mines and mineral/energy resource potential (9)
	Distance from wetlands (18)	
	Drainage (19)	
	Erosion (20)	
	Annual precipitation (23)	
	Severe weather frequency (25)	

[a]The *Excluded Areas Report* (New York State Siting Commission, 1993) states that this criterion was applied only to aboveground or belowground disposal methods during CAI and PSI. The Siting Plan, *Candidate Area Identification Report* (CAIR), and *Report on Potential Sites Identification* (ROPSI), however, indicate that this criterion was applied to all disposal methods.

underground excavations for waste storage—either in vertical shafts or horizontal "drifts." Most of the siting criteria were applied to both of these methods. A few criteria, however, were method specific (Table 3.2; see Appendix I for criteria descriptions).

Application of Screening Criteria

The first part of the screening process, Statewide Exclusionary Screening (SES), involved the application of selected exclusionary criteria to all lands in the state (see Chapter 4). Areas that failed any of the criteria used in the screening were eliminated from further consideration. The second and third parts of the screening process, CAI and PSI, involved further screening using exclusionary criteria and the quantitative scoring and ranking of nonexcluded lands using different sets of preference criteria.

The scoring system involved the application of numerical *scaling* and *weighting* factors to each criterion. Scaling factors are integer values that express numerically how well a particular area satisfies a particular criterion. With some exceptions (which are discussed later in the report), each preference criterion used in scoring was assigned a set of scaling factors in the range from 1 to 5, where 1 was least favorable and 5 was most favorable. The degree to which the preference criterion was met at a given area determined its scaling value. For example, criterion 7 preferred "areas that are distant from active or nearby abandoned mines" (see Table I.2). If a location was more than 1 mile from any mine, a scaling factor of 5 was assigned. If the closest mine was between 1/2 mile and 1 mile, a scaling factor of 3 was assigned, whereas a scaling factor of 1 was assigned if the closest mine was less than or equal to 1/2 mile away.

Weighting factors are integer values that provide a relative ranking of the numerical importance among the preference criteria in terms of site suitability—for example, the importance of distance to schools versus distance to highways. Weighting factors were developed through a process known as *values elicitation.* The Siting Commission developed the weighting factors from information obtained at a workshop attended by a small number of representatives of government, industry, and interest groups and from a similar internal Siting Commission exercise.

Weighting factors for the various preference criteria ranged from 5 to 55 (see Table I.2), and the weighting factors for all 43 preference criteria summed to 1,000 (Table 3.1). Most of the screening steps utilized only a subset of these criteria. In these cases, the weights of the criteria were renormalized so that they summed to 1,000.[4]

The Siting Plan indicates that the Siting Commission was aware of the subjective nature and importance of weighting factors. The plan states (p. 2-9):

---

[4]Except as noted in Chapter 6 for the Geographic Information System (GIS) Screening step, where a different renormalization method was used.

> In order to make comparative technical evaluations of candidate areas and sites, it will be necessary for the Commission to make value judgments about the relative importance of each siting criterion. This recognizes that some criteria are more important determinants of the overall suitability of a site than others. Quantitative expression[s] of these value judgments are called weighting factors.

In the CAI and PSI steps of the screening process, the Siting Commission used the following process to "score" each area under consideration quantitatively to assess its suitability for an LLRW disposal facility:

1. A scaling factor of 1 to 5 was assigned to each preference criterion for each area. The assignment was based on the degree to which the area satisfied the criterion based on examination of available data.
2. The scaling factor and renormalized weight for each criterion were multiplied to obtain a "score" for that criterion.
3. The scores for all of the criteria used in the screening step were added together to obtain a total score for the area. The total scores were a measure of the relative favorability of each area, with the most "favorable" areas having the highest scores. The maximum possible score for any area was 5,000.

As discussed in Chapters 5 and 6, different combinations of criteria were used in the CAI and PSI screening steps. In response to questions from the committee (Appendix F), the Siting Commission stated that the initial schedule for applying the criteria was developed by senior staff at Roy F. Weston, Inc. "based on previous experience on siting projects and the nature of anticipated data sources. . . ." As described in Chapters 5 and 6, the Siting Commission applied additional unplanned screening procedures during both of these stages.

Quality Assurance Program

The U.S. Nuclear Regulatory Commission (USNRC) and the DEC required a quality assurance program as part of site characterization and licensing efforts. The USNRC regulations regarding license applications for a low-level waste disposal site require

> a description of the quality control program for the determination of natural disposal site characteristics and for quality control during the design, construction, operation, and closure of the land disposal facility and the receipt, handling, and emplacement of waste. Audits and managerial controls must be included. (10 CFR 61.12(j))

Similarly, DEC regulations require the following:

> The Commission must submit its proposed plans for site characterization studies and a description of the proposed quality control program for such studies to the department for the department's review. . . . (6 NYCRR 382.6 (b)(3))

In addition, the report NUREG-1293, *Quality Assurance Guidance for Low-Level Radioactive Waste Disposal Facility*, (USNRC, 1989) clarifies the intent of the quality assurance requirements and provides guidance for planning the site selection and characterization process. (A full list of the supporting publications for the Siting Commission's quality assurance plan is presented in Appendix J.) From a regulatory perspective, the goal of the Siting Commission's quality assurance program was to ensure documented evidence of quality in the site selection process that would provide a basis for denying or issuing a license. A quality assurance program would stipulate procedures for a disciplined system involving planning, training, data collection, analysis, validation, and review to preserve all important information. Since New York is an agreement state, the DEC was to provide the guidance on the development and implementation of a quality assurance program.

Public Participation

As noted in Chapter 2, the 1986 State Act required the Siting Commission to keep the public informed of its activities and encourage public participation in siting and disposal method selection. In the Siting Plan the Siting Commission committed itself to meaningful public participation in all phases of its work. The following objectives were identified:

- Provide a base of public information on the LLRW disposal facility site and method selection process.
- Create convenient, meaningful opportunities for members of the public across the state to participate in those processes.
- Consider the range of public ideas and values in making the decisions that would lead to a disposal facility site and method.

The Siting Commission also committed to meet these objectives (Siting Plan, p. 1-8) by

> . . . involving the public at many points throughout the site selection. This began with the development of the site selection plan and will continue through each of the decision points leading to the selection of the preferred site.

Planned public participation programs included (1) a workshop at which representatives of state and local government, industry, and interest groups reviewed and commented on weighting factors for criteria used in site and method selection; (2) development and dissemination of public information materials, including a quarterly newsletter and fact sheets on relevant subjects; (3) a series of public meetings; and (4) solicitation of public review and comments on major documents.

The Siting Commission's stated intention for CAI screening was to recognize a range of value judgments on the relative importance of the site selection factors. Once technically suitable sites were established, the Siting Commission was to take into consideration other economic, social, and policy factors in selecting those sites upon which it would focus its efforts.

As required by the 1986 State Act, an Advisory Committee review was also to be sought. The Siting Plan describes the site selection plan workshop in which both the Advisory Committee and its invited guests had participated in August 1988. It also describes a series of meetings held in October 1988 to introduce the public to the siting process. Meetings were planned to receive input on CAI in mid-January 1989. In the spring of 1989 the meetings were to focus on identification of potentially suitable sites, and meetings planned for the summer of 1989 were to be held in candidate site communities to obtain public input on information needed to make a recommendation for certification. Input was to be sought from local citizens on the selection process, local data pertinent to the evaluations, and public values and preferences.

Throughout the process, staff were to be available to answer questions, provide presentations, and work with interest groups to provide information on the process. Local information offices were to be opened in the candidate site communities to provide a channel of communication with the Siting Commission. These offices would provide public information on radioactive waste disposal, Siting Commission plans, and the site selection process. In response to questions from this committee (Appendix F), the Siting Commission stated that approximately $5 million was expended on public participation activities between 1988 and 1993.

ANALYSIS AND DISCUSSION

Two observations are worth noting at this early point in the review of the New York siting effort. The first relates to the goals of the siting effort. As noted previously in this chapter, the stated goals for the siting process were (1) to identify potentially certifiable sites and (2) to demonstrate that no "obviously superior alternatives" could be identified readily. The second goal was interpreted by some parties in the siting process—and many members of this committee—as favoring the selection of a "best" site in some objective sense. The requirements for a site that has no obvious superior alternatives, however, are difficult to define because many factors are involved in the suitability assessment, and no one site can be expected to be superior in all respects. In any event, the Siting Commission was required only to identify certifiable sites—although, as shown in later chapters of this report, the expectations of some affected communities with regard to a "best" site created problems for the Siting Commission as siting progressed.

Second, through the implementation of the exclusionary and preference criteria, the Siting Commission divided the site selection process into an exclusionary phase, during which the commission was removing land from consideration, and a selection phase, during which the commission actively sought to identify certifiable sites from ever-smaller areas of the state. The differences between these two activities had important implications for the commission's work. The exclusionary steps, which were largely in the beginning of the process (particularly SES and the early stages of CAI; see Chapters 4 and 5), were not controversial because of their very nature. Because they excluded large regions of the state, the results of exclusionary screening were welcomed by citizens in excluded areas. By comparison, the process of selecting candidate areas and potential sites has the potential to be more visible because it focused public scrutiny on relatively small areas, and the result—being selected—was cause for concern among affected communities. In fact, as noted in Chapters 5 and 6, public interest did not become significant until the later inclusionary phases of the screening process.

By choosing a two-stage process the Siting Commission automatically, if inadvertently, heightened public scrutiny during the later stages of the screening process—where, as discussed in Chapters 5 and 6, the data were frequently mismatched to the scale of screening and the screening methodology became increasingly subjective. As later chapters show, the Siting Commission's progress slowed significantly as its focus changed from exclusion to selection. It is not clear that the commission appropriately recognized this shift of emphasis. Moreover, as far as the committee could judge, the commission failed to understand the implications of the change in emphasis from exclusion of land to selection of sites.

4
Statewide Exclusionary Screening

Statewide Exclusionary Screening (SES) was the first step in the first phase of the Siting Commission's work (Figure 3.1). This process, documented in the September 1988 *Statewide Exclusionary Screening Report* (SESR), was designed to eliminate from further consideration areas that did not meet the regulatory requirements for a low-level radioactive waste (LLRW) disposal facility. To accomplish this goal, the Siting Commission applied certain exclusionary criteria to the entire state to facilitate more detailed studies on the more promising areas.

SCREENING CRITERIA

The Siting Commission used the following five principles to select the exclusionary criteria to be applied for SES (SESR, pp. S-3 and S-4):

> • The exclusionary criteria to be applied at this stage must address conditions that are clearly prohibited in law or regulation. There can be no allowance for compensating factors.
>
> • The criteria must address only clearly defined areas with legally established boundaries. There can be no dispute or question about the data used in defining those boundaries.
>
> • The criteria must contain no qualifiers that require interpretation of the law or regulations. Such interpretation would require regulatory concurrence, which is contrary to the overall philosophy of this step that the criteria and results should be clear and not subject to any debate.
>
> • The criteria should address conditions for which data are available for all areas of the state. No

new data collection should be necessary because this is not believed to be timely or cost effective at this point in the screening process.

• The criteria should address areas that are generally large enough to appear as areas rather than points on the map and will therefore be more meaningful by reducing the area to be considered in subsequent steps.

Given these limitations, the Siting Commission selected only 5 of the 17 exclusionary criteria for the SES step. These criteria were applicable to both of the disposal methods under consideration: aboveground or belowground disposal and underground mine disposal. The specific criteria and their regulatory bases are described in Table I.1. Details of their implementation are discussed below. The Siting Commission used maps of 1:250,000 scale[1] in this screening step.

Criterion 11—Ground Water Hydrology

Exclude all areas above the Long Island Aquifer, any primary water supply aquifer, or principal aquifer designated by the Department of Environmental Conservation (DEC).[2]

Within this criterion, *primary* public water supply aquifers were defined by the DEC as "highly productive aquifers presently being utilized as sources of water by major municipal water supply systems." *Principal* aquifers were defined as "aquifers known to be highly productive or whose geology suggests abundant potential water supply, but *which are not intensively used* as sources of water supply by major municipal systems at the present time" (emphasis added).

The locations of the Long Island Aquifer and 18 primary aquifers were relatively well defined, but because only a few of the principal

---

[1]At this scale, 1 inch on the map is equal to about 4 miles on the ground (or 1 centimeter on the map is equal to 2.5 kilometers on the ground).

[2]The wording of some of the criteria varies slightly in different Siting Commission documents.

aquifers were identified, the majority were not considered at this stage. They were considered during Candidate Area Identification screening (see Chapter 5).

Criterion 32—Population Density

Exclude all villages, towns, cities, or unincorporated places, as defined in the 1980 decennial census or more recent census of the United States, New York State, or any political subdivision thereof performed by the U.S. Census, that have an average population density of more than 1,000 persons per square mile.

The wording of this exclusionary criterion matched the regulatory requirement for all but one issue: it did not make use of recent census data from New York State or smaller political entities. The intent was to ensure that comparable-quality data were applied throughout the state at this stage in the process by using the U.S. Census Bureau as the only source of information. For the screening, average population densities were calculated by dividing resident populations by the areal extent of cities and towns throughout the state. At this stage, villages were not considered because of their limited size. Also, the analysis of unincorporated areas was deferred because they lacked legally constituted civil divisions and their population densities required further interpretation.

Criterion 36—Federally Protected Lands

Exclude all lands protected by the Federal government, including: National Wildlife Refuge System; fish restoration areas; migratory bird reservations; National Wilderness Preservation System; National Wild and Scenic Rivers System; National Park System.

At the scale of statewide screening, only the lands in the National Wilderness System, National Wildlife Refuge System, and National Park

System were excluded. The smaller parcels of protected lands (e.g., migratory bird refuges) were excluded during later screening steps.

Criterion 38—State-Protected Lands

Exclude all lands protected by New York State, including components of: New York State Wild, Scenic, and Recreational Rivers System; fish restoration areas; State Park System; Adirondack Park; Catskill Park; municipal parks established as of 31 December 1987; wildlife management areas; game refuges; game farms; fish hatcheries; boat launches.

At the scale of the statewide screening, 159 state parks, including Adirondack Park and Catskill Park, were excluded. Many smaller parks and other state-protected lands were excluded during later stages of the process.

Criterion 41—Indian Lands

Exclude all Indian reservations and lands under jurisdiction of Indian nations.

Readily available data on the legal boundaries of Indian lands were used in this step of the screening process.

DATA SOURCES

Table 4.1 provides a summary of the data sources for the SES. Maps or data depicting the areas to be excluded were entered into a computerized Geographic Information System (GIS).[3] The actual polygonal shapes of the excluded features were used. The areas excluded

---

[3]A GIS is a computer system that stores geographically referenced data (e.g., map data) for manipulation, display, and analysis.

TABLE 4.1 Sources of Data Used in SES

Exclusionary Criterion (criterion number)[a]	Source of Data Used in Screening
Long Island Aquifer and primary public water supply aquifers designated by the DEC (11)	Primary aquifers digitized from New York State Department of Health maps of primary aquifers. Long Island Aquifer defined by New York State Department of Health as the political boundaries of Nassau and Suffolk counties (SESR, p. 2-11)
State parks, including Adirondack and Catskill parks (38)	New York State Department of Transportation maps and Gazetteer (SESR, p. 2-14)
National wilderness, national wildlife refuges, and national parks (36)	Sierra Club maps, 16 U.S.C. Ch. 1132 (1985, 1988), New York State Atlas & Gazetteer, New York State Department of Transportation maps (SESR, pp. 2-12, 2-14)
Cities and towns with average population densities greater than 1,000 persons per square mile (32)	1980 census data (SESR, p. 2-12)
Indian lands (41)	New York State Department of Transportation; New York Department of State

[a]See Appendix I.

by the five criteria were computer mapped and overlaid on a map of New York State. The maps of each of the excluded features were then overlaid collectively to obtain the total excluded areas.

SCREENING RESULTS

Table 4.2 summarizes the lands removed from consideration by each of the exclusionary criteria. A total of 9,669,561 acres (approximately 30 percent of the area of the state) was excluded. The locations of the excluded lands are shown in Figure 4.1. The exclusion of state-protected lands had the greatest impact compared to all other criteria throughout the siting process. The exclusion of the Adirondack and Catskill parks, and 157 additional state parks, removed over 20 percent of the state from consideration. The area excluded by population is the second largest. Among the areas excluded were New York City, Albany, Binghamton, Buffalo, Rochester, and Syracuse. Approximately 4 percent of the state was excluded using the ground water hydrology criterion; the Long Island Aquifer constituted more than 60 percent of this area.

The majority of the excluded lands were located in the southern and eastern portions of the state (the New York City metropolitan area, Catskill Park, and Adirondack Park; see Figure 4.1). A number of smaller excluded areas were distributed throughout western New York.

TABLE 4.2 Areas Excluded by SES (after Table 4-2 of the *Excluded Areas Report*, New York State Siting Commission, 1993)

Criterion (Number)	Acres Excluded	Percentage of State Excluded[a]
Ground water hydrology (11)	1,264,219	4.0
Population density (32)	1,563,142	4.9
Federally protected lands (36)	17,266[b,c]	< 0.1
State-protected lands (38)	6,737,539[b]	21.2
Indian lands (41)	87,395[b]	0.3
Total	9,669,561	30.5

[a]Percentages computed using 31,728,640 acres as the area of New York State.

[b]This acreage does not include parcels that were already excluded using other criteria.

[c]This figure was incorrectly stated as 17,250 acres in Tables S-1 and 3-8 of the SESR.

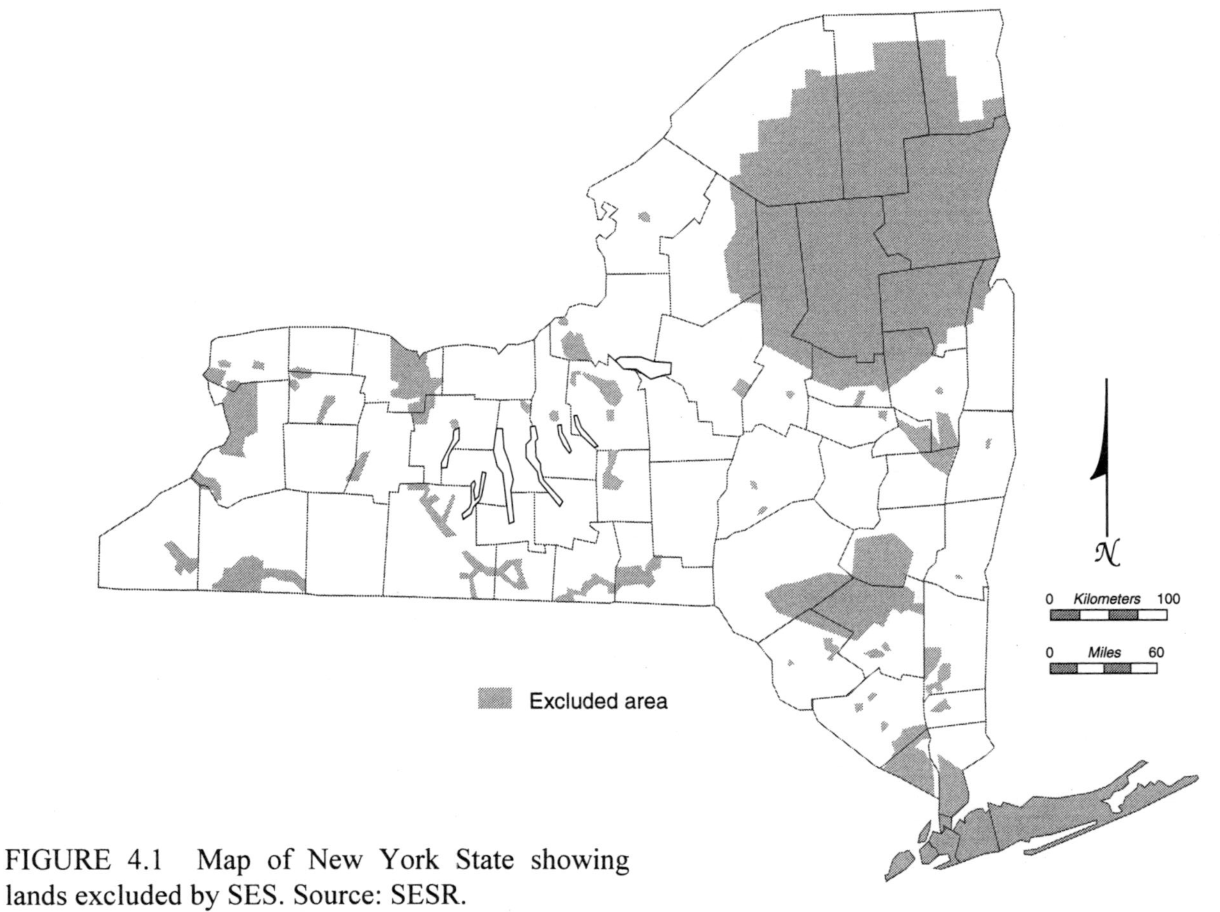

FIGURE 4.1 Map of New York State showing lands excluded by SES. Source: SESR.

PUBLIC PARTICIPATION

Members of the public did not have an opportunity for input into SES because the areas excluded during this process were mandated by law. They were, however, provided the opportunity to comment on the results at public meetings in October 1988 in Buffalo, Syracuse, White Plains, Potsdam, Albany, and Binghamton. At this early stage in the process, interest was not widespread. Attendance ranged from 13 persons in White Plains to more than 100 in Potsdam. The attendance in Potsdam was boosted by a rumor that the Siting Commission had selected St. Lawrence County as a potential site for an LLRW facility.

Informational materials available at these meetings included a program folder, five fact sheets, a program overview brochure, the summer 1988 issue of *Frontline* (the Siting Commission newsletter), the Siting Plan and executive summary, the method selection plan and executive summary, the SESR and executive summary, and the draft environmental impact statement scoping issue document. Dates and locations of meetings were advertised widely in the weeks before the meetings through public service announcements and display advertisements in newspapers.

The meetings provided a forum to inform the public about the results of the exclusionary screening, to obtain input on carrying out subsequent screening steps, and to allow input on the site and method selection plans. The public was provided an opportunity to submit written comments during a 30-day period following the meetings. The Siting Commission considered both written comments and oral presentations from the meetings in preparing its final plans. These comments were summarized in the October 1988 *Public Meeting Summary Report*. The subjects and numbers of comments included the following: site selection, 83; exclusionary screening, 35; method selection, 100; draft environmental impact statement, 13; program and schedule of the site selection process, 29; aid to local governments, 40; and miscellaneous programmatic comments, 120.

ANALYSIS AND DISCUSSION

The SES step eliminated a significant portion of the state from further consideration. In the committee's judgment, the exclusions were applied appropriately and the criteria used were based on sound regulatory considerations. The committee notes that the Siting Commission had essentially no discretion in this step of the screening process: state law or regulation precluded a disposal facility from all of the excluded areas. This portion of the selection process, the removal of land areas from further consideration, was generally seen as a positive step by observers contacted by the committee.

5
Candidate Area Identification

The purpose of this chapter is to review and analyze the Candidate Area Identification (CAI) screening process (Figure 3.1), which is described in the December 1988 Siting Commission's *Candidate Area Identification Report* (CAIR). The first part of this chapter contains a brief description of the CAI screening process, and the second part contains the committee's assessment of the process and results.

The Siting Commission's objective in this step of the screening process was to rescreen the 69.5 percent of land not excluded by the Statewide Exclusionary Screening (SES; Chapter 4) process in order to identify 10 areas for detailed investigation. Although the sizes of the candidate areas were never formally defined by the Siting Commission, it was intended that each area might be large enough to contain a number of potential repository sites. The average size of the areas selected in this process was approximately 110 square miles (approximately 260 square kilometers) that together constituted some 2 percent of the area of the state.

The Siting Commission's objectives in this step of the screening process were to

• apply exclusionary criteria that required some qualification or interpretation of regulatory requirements;

• apply exclusionary and preference criteria for conditions without strict regulatory or legal definitions of boundaries;

• apply criteria for which data were available for all areas under consideration; and

• identify those areas of the state having the greatest potential for sites for low-level radioactive waste (LLRW) disposal.

CAI actually involved three discrete screening activities, each of which involved the application of a different set of preference and (or) exclusionary criteria. The three activities were the following (Figure 5.1):

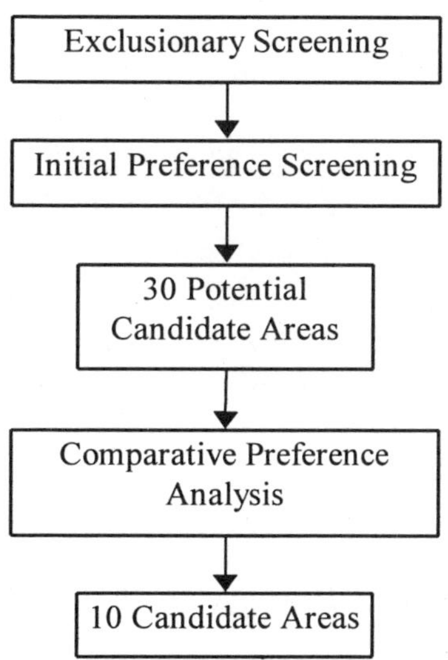

FIGURE 5.1 Flowchart for the CAI screening process.

1. *Exclusionary Screening.* Lands remaining after SES (see Chapter 4) were screened further using nine exclusionary criteria, some of which were applied at a finer scale than in SES. After this step, approximately 40 percent of the state had been removed from consideration for an LLRW disposal facility.

2. *Initial Preference Screening.* The remaining lands were scored using 14 preference criteria. A cutoff score was imposed, and the 30 highest-scoring areas were selected for further analysis.

3. *Comparative Preference Analysis.* These 30 areas were rescreened using a set of 13 preference and 5 exclusionary criteria. Ten candidate areas were selected for further study.

Each of these steps is described briefly in the following sections.

EXCLUSIONARY SCREENING

The methodology used by the Siting Commission for this Exclusionary Screening step was much like that for SES (Chapter 4). The nine criteria were applied using data from existing databases and published maps at a scale of 1:250,000 (the scale used in SES), as well as at other scales ranging from 1:500,000 to 1:24,000. Areas that did not satisfy all of the exclusionary criteria were eliminated from further consideration.

Table 5.1 lists the exclusionary criteria used in this screening process. Criterion 4 (existing mine exclusion) was applied only to sites under consideration for a mine disposal facility. Five of the exclusionary criteria (11, 32, 36, 38, and 41) had been applied during SES but at a much coarser scale. In addition, the population density exclusion (Criterion 32) was applied to all incorporated areas of the state, and all federal, state, and Indian lands were excluded (Criteria 36, 38, and 41).

The Geographic Information System (GIS) was used extensively during this and other CAI screening steps, and its analytical capabilities were utilized for the first time. The state was subdivided into cells, each having an area of 1 square mile (640 acres), for purposes of screening. At this cell size there are approximately 50,000 cells in the state, and 35,000 cells in the area of the state remaining after SES, to which the GIS was applied. If an excluded feature was contained in any part of the cell, the entire cell was removed from consideration. This screening eliminated about 9.2 percent of the area of the state and 13.2 percent of the lands remaining after SES (Table 5.1).

INITIAL PREFERENCE SCREENING

The remaining lands of the state were subjected to Initial Preference Screening in order to identify 30 potential candidate areas. The Siting Commission used the GIS to screen the nonexcluded areas of the state using the 1-square-mile grid cells employed in the Exclusionary Screening step of CAI. The following methodology was used (Figure 5.2):

TABLE 5.1 Areas Excluded by the Exclusionary Screening Step of CAI

Criterion (Number)	Acres Excluded	Percentage of State Excluded[a]
Existing mine exclusion (4)	9,600	< 0.1
Ground water hydrology[b] (11)	1,531,301	4.8
Surface water bodies (15)	1,059,840	3.3
Air quality nonattainment (26)	1,738,880	5.5
Population density[b] (32)	571,258	1.8
Federally protected lands[b] (36)	139,520	0.4
State protected lands[b] (38)	1,104,381	3.5
Indian lands[b] (41)	66,845	0.2
West Valley site[c] (46)	3,400	< 0.1
Total[d]	2,909,639	9.2

[a]Percentages computed using 31,728,640 acres as the area of the state of New York.
[b]Also applied during SES but at a coarser scale.
[c]The 1986 State Act specifically excluded West Valley from consideration as an LLRW disposal facility.
[d]Total is less than the sum because of mutually exclusive conditions.

Source: *Excluded Areas Report* (Siting Commission, 1993).

1. If the feature of interest (e.g., an aquifer) was contained in any part of the cell, the entire cell was marked as containing that feature.

2. Each cell was scored numerically for each of the 14 preference criteria (Table 5.2). A score for each criterion was obtained by multiplying the scaling factor assigned to each cell by the weight for the criterion.[1] A composite score was determined for each cell by summing the scores for all of the criteria for that cell.

---

[1]See Chapter 3 for a discussion of the numerical scoring system.

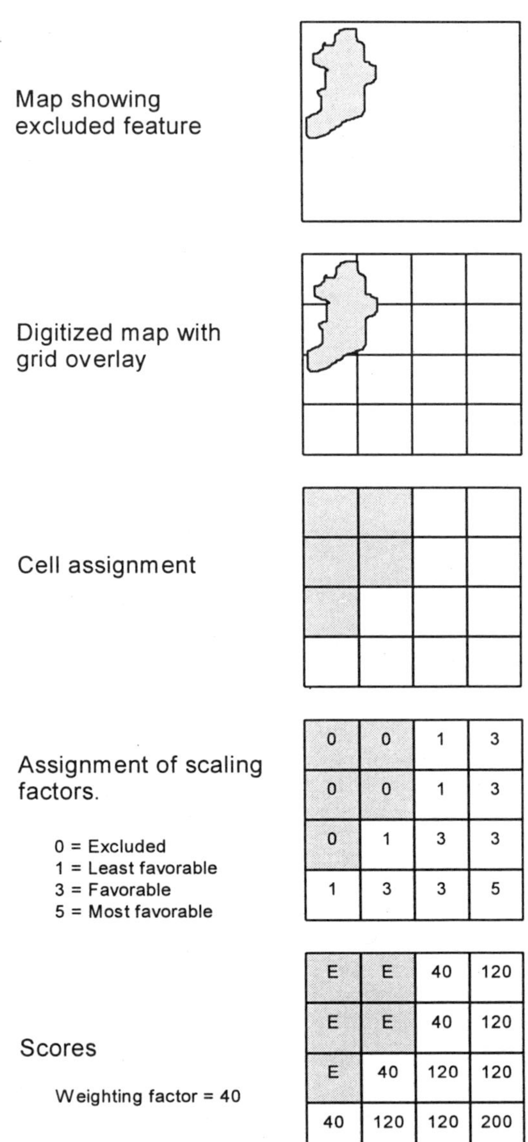

FIGURE 5.2 Conceptual illustration of scoring system using the GIS. The example shown is hypothetical, and cell size is not defined.

TABLE 5.2 Preference Criteria Applied During Initial Preference Screening

Preference Criterion	Description
2	Seismic hazards
5	Geologic units
10	Distance from oil and gas fields
12	Primary and principal aquifers
13	Surface water features
14	Unconsolidated stratigraphic units
22	Best usage of surface waters
23	Annual precipitation
24	Chronic severe weather
33	Low population densities
37	Distance from federal lands
39	Distance from state lands
42	Distance from Indian lands
51	Proximity to waste generators

The Siting Commission compared the cell composite scores in order to identify potential candidate areas. The commission found that composite scores for cells were similar throughout the majority of the towns,[2] and for this reason it compared composite scores on a town-by-town basis and defined potential candidate areas generally along town lines. Only areas containing towns with a large number of composite scores exceeding the cutoff score of 4,400 points (out of a possible 5,000 points) were considered for selection as a potential candidate area. A

[2]A town is the smallest statewide political subdivision in New York. Towns have defined boundaries and vary in size. All geographic points in the state fall within the boundaries of either a town or a city.

town whose composite scores ranged mostly between 4,400 and 4,700, and that was contiguous with another town also scoring above 4,400, could be included with the other as a single potential candidate area. A town with a large portion of cell composite scores exceeding 4,700 points could be considered a potential candidate area by itself.

A cutoff score was used because the scoring system did not produce a small number of high-scoring (i.e., clearly preferable) areas. Approximately 5 percent of the area of the state had a composite score of 4,400 or greater. Lowering the cutoff to just 4,300 points would have increased this area to about 8 percent (Figure 5.3). The Siting Commission stated (CAIR, p. 3-5) that a score of 4,400 points was chosen because it "enabled a substantial reduction of the remaining area of the State under consideration and yielded a manageable number of discrete candidate areas." As noted previously, the Siting Commission selected 30 potential candidate areas in this step.

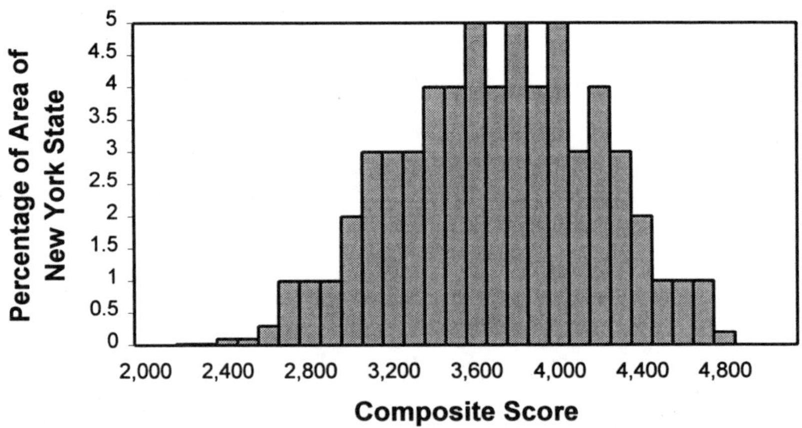

FIGURE 5.3 Distribution of composite scores for Initial Preference Screening across the area of New York State. The cutoff score was 4,400 out of a total of 5,000 points. The percentages of area of the state total approximately 60%. The other 40% of the state was excluded during previous screening steps.

COMPARATIVE PREFERENCE ANALYSIS

Comparative Preference Analysis involved the application of 5 exclusionary and 13 preference criteria (Table 5.3) to rescreen the 30 potential candidate areas. According to the Siting Plan, these criteria were not to be applied until Potential Sites Identification (PSI) screening (Chapter 6), in part because they required site-specific information. Moreover, these criteria were only a subset of the criteria to be employed during PSI screening. In applying these criteria at this earlier point in the screening process, the Siting Commission noted (CAIR, pp. 7-1, 7-2):

> Because additional criteria will be introduced in the next step of the site selection process, the Commission deemed it appropriate to seek a preview of the conditions likely to exist in the potential candidate areas for some of these additional criteria. . . . The assessments of these criteria were based on limited examinations of the areas, using readily available data sources. In some cases, the data sources were only indicators of the conditions being considered, rather than a precise measure of the actual condition. In other cases, approximations of a range of conditions had to be made, based on professional judgments of technical experts. In addition, in some cases, only the presence or absence of a feature could be noted.

Ten candidate areas were selected on the basis of qualitative screening using these criteria. For each of the criteria, the candidate areas were rated between “most favorable” and “least favorable.” The purpose of this screening was to establish the extent of the least favorable conditions and, if possible, to redraw the candidate area boundaries to exclude cells with such conditions.

The Siting Commission then assigned an “overall favorability” rating to each of the 30 potential candidate areas. The basis for these rankings was justified with the following statement (CAIR, p. 8-4):

TABLE 5.3 Criteria Applied During Comparative Preference Analysis Screening

Criterion	Description
1	Geologic complexity
3	Subsurface dissolution
4[a,b]	Existing mine exclusion
5[b]	Geologic units
6[a]	Reforestation areas
12	Primary and principal aquifers
17[a,c]	Wetlands exclusion
18[c]	Distance from wetlands
19	Drainage
20	Erosion
21	Flooding
22	Best usage of surface waters
30	Other radionuclide sources
34	Highly populated places
44[a]	Mineral soil groups
47	Transportation access
57[a,d]	Historic places
59[d]	Distance from historical/cultural resources

[a]Exclusionary criterion.
[b]Applied for new and existing mine disposal only.
[c]These criteria were applied together.
[d]These criteria were applied together.

> Based on the results of the comparative and the confirmatory evaluations, the potential candidate areas were given evaluations with regard to their overall favorability for containing technically superior sites.

All of the "more favorable" and half of the "favorable" areas were included in the list of 10 candidate areas (Figure 5.4). The process used to select these areas, particularly the processes used to select half of the favorable sites, was not well documented in the CAIR.

As part of Comparative Preference Analysis, the Siting Commission carried out a limited sensitivity analysis to examine the impact of changing preferences and weights on the selection of the 10 candidate areas. For this exercise, the 30 areas selected during Initial Preference Screening were rescored using the scenarios shown in Table 5.4. The Siting Commission found that although the scores changed somewhat for these scenarios, "most of the areas continued to display favorability scores above the cutoff level of 4,400 in each case" (CAIR, p. 5-4). Subsequently, the Siting Commission examined the change in score for each of the 1-mile-square cells within the candidate areas. For two of the scenarios (Scenarios 4 and 5 involving Criterion 51, proximity to waste generators), cells in five candidate areas fell below the cutoff. Despite this result, the Siting Commission decided not to change the screening results. It concluded (CAIR, p. 5-4):

> This factor [Criterion 51—proximity to waste generators] is not deemed to be of overriding significance, considering that New York waste generators are currently shipping their wastes to much more distant disposal sites out of State. The consistency of results from the sensitivity studies provides assurance of the reasonableness of the weighting factors used by the Commission in identifying the potential candidate areas.

Notably, 3 of the areas that fell below the cutoff in this sensitivity analysis were among the 10 candidate areas, and sites from 2 of these areas were selected in the next step of the screening process (Chapter 6).

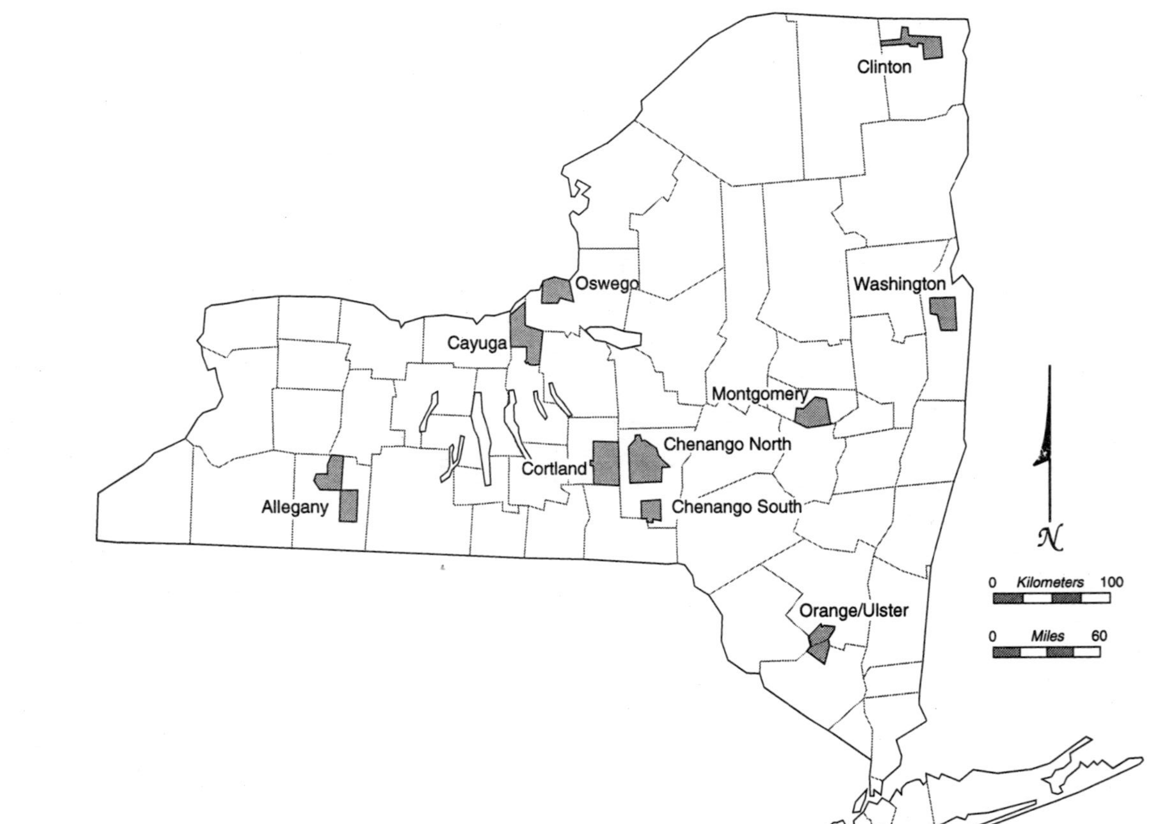

FIGURE 5.4 Map of New York State showing the 10 candidate areas. Source: CAIR.

TABLE 5.4 Scenarios Used for Sensitivity Analyses During Comparative Preference Analysis Screening (from CAIR, pp. 5-3 and 5-4)

Case Number	Scenario (criterion number)
1	Base case, no change
2	Remove annual precipitation and chronic severe weather criteria (23 and 24)
3	Remove proximity to waste generators criterion (51)
4	Double the weight of the proximity to waste generators criterion (51)
5	Triple the weight of the proximity to waste generators criterion (51)
6	Double the weights for the criteria regarding distance from federal, state, and Indian lands (37, 39, 42)
7	Double the weight for distance from oil and gas fields criterion (10)
8	Double the weight for the low population densities criterion (33)
9	Increase the weights of the water resources criteria by 50% (12, 13, 22)
10	Increase the weight of the geologic criteria by 50% (2, 5, 14)
11	Delete all criteria except 5, 12, 14, and 33
12	Delete all criteria except 5, 12, 13, 14, 22, and 33

PUBLIC PARTICIPATION

CAI screening marked the transition from the exclusionary to the selection phase of the New York siting effort, although the Siting Commission did not make that distinction. Correspondingly, the relationship between the public and the Siting Commission changed significantly. The public was largely unaware of the Siting Commission's progress and was unprepared for the release of the CAIR. Announcement of the 10 candidate areas raised public awareness in the affected communities and marked the beginning of a concerted effort by some to oppose the siting process. In the meetings following this announcement, the Siting Commission was unprepared for the intensity of public scrutiny of its siting decisions.

Following the release of the CAIR in December 1988, only one month after the release of the final version of the Siting Plan (see Table 2.1), the Siting Commission held public meetings to explain the process of selecting a site and a disposal method and to provide the opportunity for public comment. All of the counties with candidate areas contended that none of their land was suitable for an LLRW disposal facility. Comments from the counties focused on perceptions that the process was unfair and unscientific, that it was based on inadequate or inaccurate information, and that it involved subjective judgments. The sentiment of Oswego County in its response to a Siting Commission request for technical information reflected the perceptions of some that the screening process was designed to identify the "best" site for an LLRW disposal facility: "We do not accept that we are one of the ten best sites in the state."

In early 1989 the New York State Department of Health conducted workshops on the siting process for local health officials from the 10 candidate areas. The department also met with local officials and interested individuals in each of the candidate areas to discuss its public information program and upcoming meetings on the health effects of radiation. Attendance at these meetings ranged from 100 to 900 persons, and, as a result, the Siting Commission's mailing list grew to approximately 10,000 individuals.

Difficulties in dealing with the public were becoming apparent. Public comments were consuming more time at Siting Commission meetings—almost three hours at the February 1989 meeting alone. At the

March 1989 meeting of the Siting Commission, the public was asked to submit questions on 3 x 5 cards; no follow-up questions were accepted because of time limitations and the number of questions. This policy was not well received by meeting participants.

The Siting Commission also came under criticism for its delays in responding to public comments. At its March meeting, the commission attributed delays in responding to public comments to a vacancy on the secretarial staff, and it was reported that a staff liaison program and issues tracking system had been instituted.

In June 1989 the Siting Commission began videotaping its monthly meetings. Videotapes of the public meetings along with other informational materials were mailed to local libraries. These tapes were viewed by members of this committee as part of the review process.

ANALYSIS AND DISCUSSION

The committee identified problems with the design or application of several of the criteria used in CAI screening. The comments below address only those criteria used in Exclusionary Screening and Initial Preference Screening. Comments on criteria used in Comparative Preference Analysis screening are deferred to Chapter 6.

Criterion 2—Seismic Hazards

> *Prefer areas of lower seismic hazard based on predicted maximum horizontal ground acceleration as given by Algermissen et al. (1982).*

This was designed to be a preference criterion in the Siting Plan, but the committee believes that there should have been a related exclusionary criterion in addition to this preference criterion. The New York State Department of Environmental Conservation (DEC) regulations state:

> Areas must be avoided where tectonic processes such as faulting, folding, seismic activity, or vulcanism may

> occur with such frequency and extent as to affect the ability of the disposal site to meet the objectives of Subpart C of this Part, or as to preclude defensible modeling and prediction of long-term impacts. (6 NYCRR 382.21(e))

The Siting Plan justifies the omission of an exclusionary criterion for seismic hazards with the statement, "No separate criterion is presented for Quaternary (active) faults because in New York such faults, if they exist, are encompassed by those areas predicted to have higher probabilities of seismic activity" (p. A-12). This statement assumes that current seismological and geological investigations have identified the active faults in New York State and that those active faults are reflected in the Algermissen et al. (1982) seismic hazard map, which was used by the Siting Commission to define areas of seismic hazard. In the committee's judgment, neither of these assumptions is valid.

Although earthquakes do occur in New York, studies to date have failed to identify positively any faults that are known to be seismically active. The Algermissen et al. (1982) map used broad seismic zones, based on seismicity and regional geology in New York State, for computation of probabilistic ground accelerations. Active faults will not necessarily lie in the areas of high probabilistic ground accelerations. Some active faults, such as parts of the San Andreas Fault in California, have low levels of seismicity but are known to have the potential for very large earthquakes.

There are two primary earthquake hazards to a low-level radioactive waste facility. The first is the possibility of surface faulting breaching the facility. Such faulting can damage the facility and provide a direct geologic path for contaminant migration. It is complicated and costly to engineer against such an eventuality, so active faults should be avoided in site selection. The second hazard is shaking from an earthquake that could damage the structural integrity of the facility. The level of ground shaking necessary to damage a facility depends upon a number of factors, including the type of facility, its design and engineering, and the duration of ground shaking.

Even though no seismically active faults have been identified in New York State, the committee believes there should have been an exclusionary criterion for active faulting for the following two reasons: (1) In designing the exclusionary criteria, the Siting Commission could not know in advance whether or not a strong earthquake with surface faulting would occur, so the only prudent course is to cover that possibility; and (2) the Siting Commission needed to have a way of dealing with the emergence of geologic information on active faulting once the criteria were set. For example, an exclusionary criterion would have disqualified a site if a county brought forward evidence that there was an active fault at one of the sites under consideration or if the Siting Commission discovered a recent fault during on-site investigations.

The committee also believes that the current seismic hazard preference criterion (i.e., Criterion 2) should be retained as such. This criterion prefers areas in the state with the lowest expected earthquake ground motions that would likely be experienced by a facility during its design lifetime. The application of this criterion minimizes the chances that a facility will experience unexpectedly strong ground motions against which it was not engineered.

Criterion 4—Existing Mine Exclusion

Exclude all abandoned mines and all geologic units that are less than or equal to 30 meters below the ground surface.

This criterion applies only to mine disposal sites. It is derived from 6 NYCRR 382.35(b)(1),[3] which calls for a mined repository to be placed more than 30 meters below the ground surface. In siting a mine disposal facility, the criterion excludes (1) existing mines that are less than 30 meters below ground[4] and (2) geologic host units that are less than 30 meters below ground. During CAI Exclusionary Screening, the Siting Commission implemented the first exclusion.

The Siting Commission applied this exclusion using the following procedure. The depths and locations of mines that had not been excluded by other criteria were entered into the GIS, and cells including

[3]Title 6, Part 382, *New York Code of Rules and Regulations.*

[4]And thus excludes surface quarries from consideration for the mine disposal method.

mines less than 30 meters deep were excluded. The regions surrounding the mines were then scored using the preference criteria that were applicable to a mined disposal facility. With a cutoff score of 4,400 points, there were 25 existing mines under consideration, most of which were located outside the 10 candidate areas.

The Siting Commission implemented the second exclusion during the Comparative Preference Analysis stage of CAI as part of the evaluation of candidate areas for new mines. At this time, potential candidate areas that contained no potentially suitable geologic units (Paleozoic carbonates, Paleozoic salt, or Precambrian metamorphic rocks) at greater than 30 meters depth were excluded from hosting a new mine.

The official status of existing mines located at depths greater than 30 meters was unclear at the end of CAI screening. The report (CAIR, p. 7-19) notes that in the evaluation of preference Criterion 1 (geologic complexity),

> all of the mines fell into in [*sic*] the least favorable category because they are located in geologic settings with relatively complex stratigraphic and/or structural features, and are therefore considered as least favorable for a potential existing mine site.

The report makes no statements, however, regarding exclusions of the mines or whether they were to be carried forward for further evaluation. At a meeting between the Siting Commission and Weston, Inc., staff, all existing mines were officially excluded because of "geologic complexity," in effect broadening the scope of Criterion 4. At the time, this was a controversial decision because the possibility of constructing a new mine to host a disposal facility was still being considered.

In the committee's judgment, there are sound technical reasons to broaden Criterion 4 to exclude all existing mines. "Geologic complexity," however, is one of the weaker justifications. The Siting Commission had insufficient information to assess this criterion for individual existing mines and even less for the siting of new mines. As discussed in Chapter 6, the committee believes that the "geologic complexity" criterion had limited value in the site selection process.

The committee believes that the following technical reasons could have been used to exclude existing mines, regardless of depth. First, abandoned mines are frequently in disrepair, they are subject to collapse, and they frequently experience flooding. Second, many abandoned and existing mines have limited-access conditions. Finally, sites over existing mines may be subject to subsidence with resultant damage to surface facilities. In general, these problems can be traced to the fact that abandoned and existing mines are designed primarily for resource extraction and secondarily for structural stability and hydrologic isolation.

Criteria 11 and 12—Protection of Ground Water Resources

Criterion 11 excludes both primary and principal aquifers designated by the DEC, specifically:

> *Exclude all areas above the Long Island Aquifer, any primary water supply aquifer, or principal aquifer designated by the DEC.*

The first two components of this criterion (exclude Long Island and primary aquifers) were implemented during SES (Chapter 4). The third (exclude principal aquifers) was carried out during CAI screening. Maintaining a distance from primary and principal aquifers is also listed as a preference in Criterion 12:

> *Prefer areas that are distant from primary or principal aquifers, hydrogeologic units that fit the definition of principal aquifers, and well head areas for community water supply systems.*

Both of these criteria hinged on the DEC's definition (and designation) of principal and primary water supply aquifers. To this end, a primary aquifer was defined as

> . . . a highly productive water bearing formation identified by the department consisting of unconsolidated

> (non-bedrock) geologic deposits, which: (1) receives substantial recharge from the overlying land surface; and (2) is presently utilized as a major source of water for public water supply. . . . (6 NYCRR 382.2(tt))

By comparison, principal aquifers referred to

> . . . unconsolidated (non-bedrock) geologic deposits identified by the department which: (1) receive substantial recharge from the overlying land surface; (2) are known to be highly productive or whose geology suggests a potentially abundant source of water; and (3) are not presently used as a major source of water for public water supply. (6 NYCRR 382.2(uu))

The Siting Commission had no discretion in the application of Criterion 11, because it was specified by regulation. Moreover, the commission was required to perform the screening using the primary and principal aquifers identified by the DEC.

Criterion 12, on the other hand, allowed the Siting Commission to select areas distant from any hydrogeologic unit that fit the definition of principal aquifers,[5] not just principal aquifers that had been identified previously by the DEC. During CAI screening, many of the counties hosting candidate areas provided evidence to the Siting Commission that their ground water resources were comparable to other principal aquifers in New York State. There were no provisions in the Siting Plan for the Siting Commission to examine these data at this stage of screening. The committee believes there should have been provisions to accommodate contributed data on aquifer characteristics, well performance, and water usage on the scale of the 1-mile-square cells used in CAI screening,

---

[5]It should be noted that the DEC definitions of primary and principal aquifers do not include aquifers that may be found in fractured bedrock (e.g., granite, shale, or sandstone), even though these may be locally important water supplies for farms and small communities.

particularly in light of the fact that the Siting Commission used outside data to reclassify two aquifers in the Siting Plan.[6]

Criteria 12, 13, and 18—Buffers from Water Resources

These performance-oriented criteria express a preference for sites that are greater than 1 mile from a primary or principal aquifer, a significant surface water feature, or a wetland. In response to committee questions on this issue (Appendix F), the Siting Commission indicated that the 1-mile standard was based on the assumption that ground water flow rates would average about 10 feet per year. Consequently, about 500 years would be required for ground water to travel 1 mile, which would satisfy the structural stability requirements for Class C wastes. The Siting Commission also indicated that the scaling for these and other distance criteria was based largely on qualitative assessments:

> All of the scaling factors are intended to simply guide the site screening process and none is an absolute measure of the performance of a site in protecting public health and safety.

The committee acknowledges the desirability of buffers to protect water resources. The committee notes, however, that to be most effective, such buffers must be based on local flow conditions, which depend on factors such as slope, soil type, subsurface structure, and geology. Such data were not available to the Siting Commission for all parts of the state at this stage of screening; consequently, the evaluation of these criteria should have been deferred to a later stage in the screening process.

Criterion 22—Best Usage of Surface Waters

> *Prefer areas where location of a disposal facility is unlikely to impair the best usage of surface waters, as measured against water quality standards.*

---

[6]Based on public comments on the draft Siting Plan, the Siting Commission reviewed data on other aquifers and classified the aquifers at Tug Hill and Cattaraugas Creek as equivalent to principal aquifers.

Criterion 22 defines the favorability of a site relative to the current or future use of surface waters, particularly how that usage might be affected by siting an LLRW disposal facility. Scaling factors were assigned based on numerical buffer zones around surface water intakes for community water systems. For example, for intakes from water bodies less than 20 square miles in area, a site falling outside a 2-mile-wide buffer zone surrounding the entire water body would be assigned a scaling factor of 3.

As noted above, the committee acknowledges the desirability of buffers to protect water resources. To be most effective, however, such buffers must be based on local flow conditions, which depend on site-specific conditions. Because such data were not available to the Siting Commission for all parts of the state at this stage of screening, the evaluation of this criterion should have been deferred to a later stage in the screening process.

Criterion 26—Air Quality Nonattainment

Exclude areas that are within pollutant nonattainment areas.

The committee questions whether this exclusionary criterion is necessary because it appears to have no basis in legislation or existing regulation. The Siting Plan notes that development of a facility in an air quality nonattainment area could preclude the future use of on-site incineration as a waste processing method. While the ability to locate an incinerator at the disposal site would provide flexibility in the waste management system, in the committee's judgment it is not a *necessary* condition for an LLRW disposal facility. The committee notes that air quality could be affected during construction of the facility but that engineering controls could be utilized to minimize adverse impacts. Indeed, construction activities are conducted routinely in air quality nonattainment areas.

The committee also notes that because air quality nonattainment areas do not generally coincide with rural areas, this criterion is inherently biased against rural areas (i.e., rural areas are less likely to be excluded on the basis of this criterion). Most of the land excluded by this criterion overlapped with areas already excluded by SES; however, the criterion did exclude an additional 5 percent of the state. Very little of this land was rural. In the committee's judgment, if the criterion was to be included at all, it should have been included as a preference criterion.

Initial and Comparative Preference Screening

As noted previously, the Siting Commission chose a cutoff score of 4,400 points in order to select the 30 candidate areas during Initial Preference Screening. In the committee's judgment, there are two problems with the Siting Commission's use of arbitrary cutoff scores to select a small number of sites from a relatively large area of the state. First, as will be shown by the following sensitivity analysis, there is not a good correlation between the numerical score and the likely performance[7] of a site as an LLRW repository, owing to the way in which exclusionary and preference criteria were applied in the screening process. Thus, potentially suitable sites may have been screened out prematurely, and less suitable sites may have been retained. Second, problems were also created by the decision to use the criteria to screen out all but a few sites in the state. Even if there were a reasonably good overall correlation between numerical score and site suitability, the correlation across a truncated range of scores (e.g., over a range of a few hundred points, say from 4,000 to 4,500 points) may be poor, and there may be little if any correlation within the highly truncated range of scores when only the few top-scoring sites are considered.

The 10 candidate areas identified through Comparative Preference Analysis screening were not the highest-scoring areas

---

[7]The Siting Commission's objective in screening the state was to select a site that would meet the requirements of federal and state regulations for a land disposal facility. The emphasis of the regulations is on selecting a site capable of preventing the off-site movement of radionuclides released from an LLRW disposal facility. Since the primary release pathway is through the surface and shallow subsurface environment, state regulations focus on the geologic and hydrologic factors that control the movement of radionuclides introduced into these settings. The committee uses the term *performance* to describe the degree to which these geologic and hydrologic factors contribute to the integrity and stability of the repository and the prevention of radionuclide release to the accessible environment.

identified during Initial Preference Screening.[8] Indeed, it is unlikely that the same 10 candidate areas would have been selected had the entire set of 43 preference criteria (see Appendix I) been applied to all of the nonexcluded lands. In other words, the screening results are not unique—the areas selected depend on the sequence of screening steps and the mix of criteria applied at each step. This is a fundamental shortcoming of the screening process.

Although a limited sensitivity analysis was undertaken by the Siting Commission, it was poorly documented in the CAIR, and it had a number of serious shortcomings. The analysis tested the sensitivity of scoring only for the 30 areas selected during Initial Preference Screening. It did not address whether scores for other areas of the state would change significantly if the criteria were applied in a different order. It appears to the committee that the sensitivity analysis may have been used as a means to justify the candidate areas selected, rather than as an unbiased effort to evaluate the process prior to its application.

The limitations of the Siting Commission's sensitivity analysis can be demonstrated through a simple exercise that compares scores for several hypothetical scenarios for Initial Preference Screening (Table 5.5). Despite its simplicity, this exercise provides several components of a sensitivity analysis that would have been extremely valuable in planning the screening process.

Table 5.5 divides the 14 preference criteria used in Initial Preference Screening into three groups:

1. Preference criteria based on *known quantities*—readily available data having relatively uniform quality across the state. Such data include average annual precipitation; locations of federal, state, and Indian lands; proximity to waste generators; and distance from surface water features. Most of these criteria address nonperformance factors. That is, they are not critical issues in determining the capability of a site to function as an LLRW disposal site.

2. Preference criteria based on *uncertain quantities*—data of unknown or uneven accuracy or completeness. Most of these preference criteria address the likely performance of the site as a host for an LLRW repository. These include geology, proximity to aquifers, best usage of

[8]The 10 candidate areas ranked 1, 8, 9, 10, 11, 20, 21, 24, 25, and 27 out of 30.

surface waters, distance to active and abandoned oil and gas fields, and the probability of seismic activity. In the scoring system used by the Siting Commission (Chapter 3), these criteria tend to give high scores to areas that lack data.

3. Preference criteria based on *population quantities*—in this case, demographic information. These quantities can be measured readily, but they are sensitive to such details as placement of boundaries and annexation.

The committee applied these three sets of preference criteria to seven hypothetical scenarios to test the effects of various scaling schemes; the results are shown at the bottom of Table 5.5. Scenario 1 represents the "base case" in which all of the preference criteria are assigned a scaling factor of unity (least favorable). The total score[9] is 1,000 points, the minimum score an area can receive. The effect of population density on scoring is isolated in Scenario 2. Scenario 3 provides a test of the sensitivity of scoring to the presence of 1-mile buffer zones for federal, state, and Indian lands. Scenarios 4, 5, 6, and 7 assess the influence of preference criteria scored with "uncertain" data, as explained in the footnotes in Table 5.5.

A significant result of this sensitivity analysis is that none of the scenarios give scores that exceed the cutoff value of 4,400 points. This is a positive attribute of the screening process because it indicates that favorable scores for a large number of criteria are required to qualify a site.

The high scores shown in Table 5.5 for Scenarios 4, 5, 6, and 7 indicate that the preference criteria analyzed using uncertain data exert a strong control on scoring. Scenario 4 demonstrates, for example, that about 2,700 of the necessary 4,400 points can be attributed to highly favorable scores on these criteria alone. Because these criteria address performance factors for an LLRW facility, an area's scores may be

---

[9]As noted in Chapter 3, the total score is equal to the sum of the scaling factors for each criterion multiplied by their respective weights.

TABLE 5.5 Initial Preference Screening Sensitivity Analysis

Category (Criterion Number)	Criterion Weight[a]	Scaling Factor[a,b]						
		Scenario 1	Scenario 2	Scenario 3	Scenario 4	Scenario 5	Scenario 6	Scenario 7
Known Quantities								
Surface water features (13)	107	1	1	5	1	1	5	5
Annual precipitation (23)	53	1	1	1	1	1	1	1
Chronic severe weather (24)	53	1	1	1	1	1	1	1
Distance from federal lands (37)	27	1	1	5	1	1	5	5
Distance from state lands (39)	27	1	1	5	1	1	5	5
Distance from Indian lands (42)	27	1	1	5	1	1	5	5
Proximity to waste generators (51)	53	1	1	1	1	1	1	1
Subtotal	347							
Uncertain Quantities								
Geology (5, 14)	200	1	1	1	5	5	5	5
Distance from oil and gas fields (10)	53	1	1	1	5	5	5	5

Primary and principal aquifers (12)	147	1	1	1	5	5	5	5
Best usage of surface waters (22)	80	1	1	1	5	5	5	5
Seismic hazards (2)	53	1	1	1	5	5	5	5
Subtotal	533							
Population Quantities								
Low population densities (33)	120	1	5	5	1	5	1	5
Subtotal	120							
Total	1,000	1,000	1,480	2,232	3,132	3,612	3,884	4,364

[a]See Chapter 3 for an explanation of these quantities.

[b]Scenario 1: Least favorable [scaling factor (sf) = 1] for all criteria.
Scenario 2: Most favorable (sf = 5) for low population densities.
Scenario 3: Most favorable (sf = 5) for low population densities, distance from federal, state, and Indian lands, and surface water features.
Scenario 4: Most favorable (sf = 5) for all "uncertain quantities."
Scenario 5: Most favorable (sf = 5) for all "uncertain quantities" and low population densities.
Scenario 6: Most favorable (sf = 5) for all "uncertain quantities," distance from federal, state, and Indian lands, and surface water features.
Scenario 7: Most favorable (sf = 5) for all "uncertain quantities," low population densities, distance from federal, state, and Indian lands, and surface water features.

correlated weakly with the potential performance of a facility located in that area.

The results for Scenario 7 illustrate the influence of a poorly designed criterion on scoring. The area in Scenario 7 had the most favorable performance and population scores, yet it scored just below the cutoff. If the annual precipitation for this area was decreased by a small amount (e.g., from 50 to 49 inches per year), the score for Criterion 23 would have increased by 106 points, and the total score would have exceeded the cutoff. This example suggests that insignificant performance or licensing factors could have played a significant role in the selection of candidate areas, and it further supports the committee's conclusion that preference scores are very subjective and may not have been a strong indicator of technical suitability of a site for an LLRW disposal facility.

SUMMARY

The focus of this chapter is the CAI step of the screening process. The objective of this step was to screen the 69.5 percent of state land remaining after SES to identify 10 candidate areas, each with an average size of about 110 square miles (approximately 260 square kilometers), that together constituted about 2 percent of the area of the state.

The committee analyzed the exclusionary and preference criteria used in Exclusionary Screening and Initial Preference Screening (Tables 5.1, 5.2) and the screening process itself. The committee's findings are summarized in the following sections.

Screening Criteria

Various combinations of the exclusionary and preference criteria were applied during CAI screening. The committee identified problems with several of these criteria, as noted below (see also Chapter 6):

• *Criterion 2—seismic hazards*—should have included an exclusionary criterion for active faulting through a site in addition to the preference criterion for areas of lower seismic hazard based on the probabilistic seismic hazard maps. This exclusionary criterion should have been applied during the site studies phase of screening (see Figure 3.1).

• The scope of *Criterion 4—existing mine exclusion*—was essentially broadened by the Siting Commission to exclude all existing mines, regardless of their depth below the surface. Although there are sound technical reasons to exclude existing mines, due to their inadequate design for an LLRW disposal facility, the Siting Commission's justification—geologic complexity—could not be properly evaluated using data available during CAI screening.

• *Criteria 11 and 12—protection of ground water resources*—The application of Criterion 11 was consistent with DEC regulations. In applying Criterion 12, however, the Siting Commission should have made provisions to accommodate contributed data on aquifer characteristics, well performance, and water usage on the scale of the 1-mile-square cells used in CAI screening.

• *Criteria 12, 13, and 18—buffers from water resources—and Criterion 22—best usage of surface waters*—The committee notes that buffers are desirable but should be based on local flow conditions, data for which were not available for all parts of the state at this stage of screening. The evaluation of these criteria should have been deferred to a later stage in the screening process.

• *Criterion 26—air quality nonattainment*—appears to have no basis in legislation or regulation and should have been designated as a preference criterion.

Screening Process

CAI screening comprised the following three discrete activities: (1) Exclusionary Screening, (2) Initial Preference Screening, and (3) Comparative Preference Analysis. The committee identified problems with the following aspects of the screening process:

• The Siting Commission's use of an arbitrary cutoff score for Initial Preference Screening is not justified because there is not a strong correlation between the numerical score and the likely performance of the area as an LLRW repository, especially when only a few sites are selected from a large area of the state.

- The results of screening are nonunique: the candidate areas selected depend to a great extent on the sequence of screening steps and the combination of criteria applied at each step, rather than the likely performance of the site.
- The sensitivity analysis performed by the Siting Commission was documented poorly and had a number of technical shortcomings. The analysis was incomplete, and it was applied to only 30 areas of the state.

The committee performed its own sensitivity analysis of the 14 preference criteria used in Initial Preference Screening. Based on this analysis, the committee made the following observations:

1. A site must be favorable for a relatively large number of criteria in order to exceed the cutoff score, which is a positive attribute of the screening process.
2. Preference criteria scored using data of unknown or uneven accuracy and completeness exerted a strong control on scoring and biased results toward favoring regions that lacked data. Because these criteria address performance factors for an LLRW repository, and because the data are so uncertain, the site scores may be correlated only weakly with the potential performance of the sites.

6
Potential Sites Identification

The Potential Sites Identification (PSI) step of the screening process (Figure 3.1) was described in the September 1989 Siting Commission's *Report on Potential Sites Identification* (ROPSI), which was published about 10 months after the Siting Plan itself (see Table 2.1). The report was published as part of New York State's effort to meet the January 1, 1990, milestone set by the 1985 Amendments Act (Table 2.1) so that waste generators in New York could retain access to disposal sites in other states. The objective of this screening step was to identify a small number of *potential sites* for a low-level radioactive waste (LLRW) disposal facility from among the 10 candidate areas identified in Candidate Area Identification (CAI) screening (Chapter 5). The PSI process involved detailed screening of the candidate areas, limited field observations at individual sites, and comparative evaluations. It resulted in the selection of five potential sites (Figure 6.1).

PSI screening involved four discrete activities, each involving the application of a different combination of exclusionary and preference criteria (Figure 6.2):

- *Geographic Information Systems (GIS) Screening.* The 10 candidate areas were screened using 13 exclusionary and 27 preference criteria to identify 96 sites, which ranged in area from one-half to several square miles.
- *Qualitative Map Assessments.* These 96 sites were rescreened using 1 exclusionary and 5 preference criteria. A total of 51 sites were selected.
- *Field Surveys.* Limited field inspections ("windshield surveys") were performed on these 51 sites and 4 other "offered" sites[1] to identify unfavorable conditions. A total of 19 sites were selected for further review.

---

[1]These sites were offered to the Siting Commission as possible host sites for an LLRW disposal facility by the landowners. These sites are discussed in more detail later in this chapter.

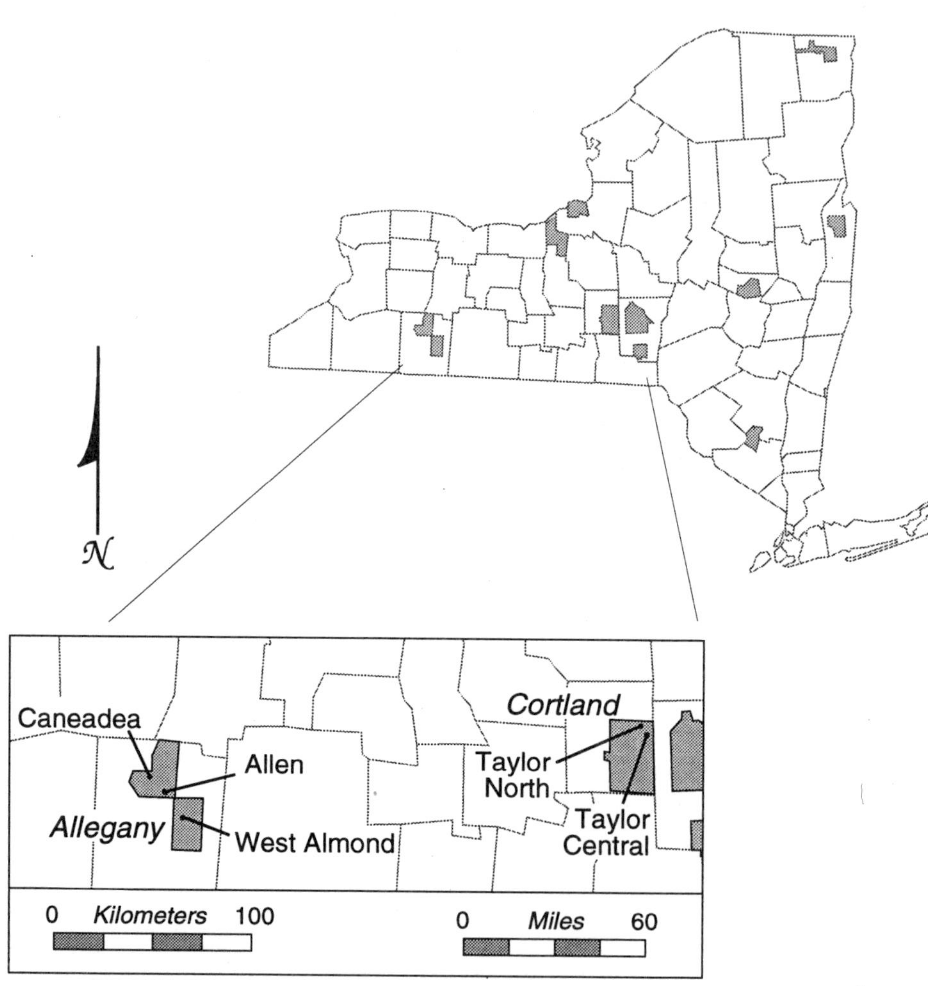

FIGURE 6.1 Map of New York State showing the five potential sites, with the candidate areas included for reference. Source: ROPSI.

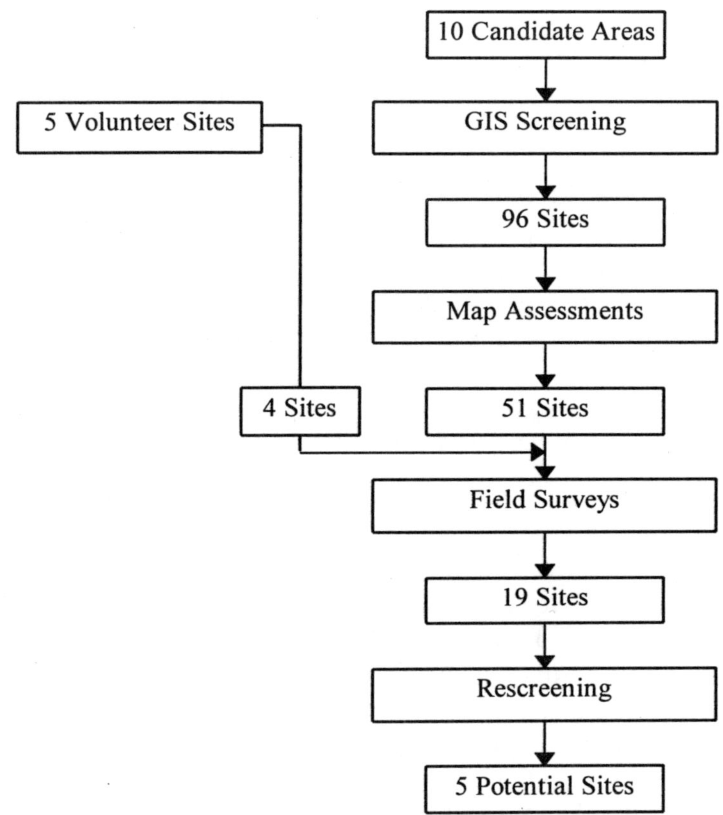

FIGURE 6.2 PSI screening process flowchart.

• *Rescreening Using All Criteria.* These 19 sites were rescreened using all 17 exclusionary and 43 preference criteria defined in the original siting plan (Appendix I). Five potential sites were selected.

Table 6.1 shows the land areas of the state that were eliminated using the exclusionary criteria in the various steps of PSI screening.

Shortly before the Siting Commission issued the ROPSI, it issued the *Disposal Method Screening Report* (1989a), which ruled out the use of vertical shaft mines for disposal. Consequently, drift mine disposal was the only underground storage method under consideration during PSI. At this time, the Siting Commission screened areas and sites for both aboveground or belowground methods and for drift mine disposal.

GIS SCREENING

GIS Screening was carried out at a scale of 1:24,000,[2] which represented about a 10-fold increase in resolution from CAI screening. A grid consisting of 40-acre areas, or cells, was imposed on each of the 10 candidate areas identified through CAI screening. Each of the cells was evaluated using all of the exclusionary and preference criteria applied during CAI screening, in combination with an additional set of two exclusionary and six preference criteria. A list of these criteria is given in Appendix I. It is important to note that this group of criteria was significantly smaller than the full set specified in the original Siting Plan. Many of the socioeconomic criteria and a few of the performance criteria were not applied in this step of screening.

As in previous screening steps, cells containing excluded features were eliminated from consideration. The remaining cells were then scored using the preference criteria.[3] Based on the distribution of scores

[2]At this scale, 1 inch on the map is equal to 2,000 feet on the ground (or 1 centimeter on the map is equal to 240 meters on the ground).

[3]In this screening step, scores were calculated in a different way than in previous steps. In particular, the weights of the preference criteria were not normalized to 1,000 before composite scores were calculated. Instead, composite scores were calculated using the "raw" weights shown in Table I.2, and the composite scores were then normalized to

TABLE 6.1 Lands Excluded During PSI Screening (after New York State Siting Commission, 1993, p. 4-6, table)

	Acres Excluded		Percentage of State Excluded[a]	
Criterion	Aboveground/ Belowground	Mines	Aboveground/ Belowground	Mines
6	105,100	105,100	0.33	0.33
11	2,120	2,120	< 0.01	< 0.01
15	12,320	—	0.04	—
16	169,200	169,200	0.53	0.53
17	140,680	—	0.44	—
28	5,840	5,840	0.02	0.02
32	2,320	2,320	< 0.01	< 0.01
36	3,800	3,800	0.01	0.01
38	26,720	26,720	0.08	0.08
41	600	600	< 0.01	< 0.01
44	170,300	170,300	0.54	0.54
57	1,280	1,280	< 0.01	< 0.01
Total[b]	463,360	404,800	1.46	1.28

[a]Percentages computed using 31,728,640 acres as the area of the state of New York.
[b]Total is less than the sum because of mutually exclusive conditions.

and the need to select a manageable number of sites, the Siting Commission imposed a cutoff score of 3,900 points over a minimum of five contiguous 40-acre cells in an approximately square pattern. A total of 96 sites with scores of 3,900 or higher were identified.

5,000 points. Except for rounding differences, this method should yield results identical to those obtained by normalizing the weights.

QUALITATIVE MAP ASSESSMENTS

The 96 sites were rescreened qualitatively using criteria that could be evaluated on the basis of map data (Table 6.2). For each of these criteria, the sites were assigned qualitative scores of "+," "0," or "–." These scores were then aggregated, and an overall rating was applied as shown below:

A: A site received no more than one 0 score and no – scores.
B: A site received more than one 0 score and no – scores.
C: A site received at least one – score.

Sites that received an A or B rating and that were at least 400 acres in size were selected for further analysis.

FIELD SURVEYS

The 51 sites selected by Map Assessments screening, and 4 additional sites offered by private landowners (see discussion later in this chapter), were subjected to reconnaissance, or "windshield surveys."[4] The purpose of these surveys was to identify changes in the sites since

TABLE 6.2 Criteria Used for Qualitative Map Screening

Criterion Number	Description
31, 35	Proximity to incompatible activities/ nonresident populations
44[a]	Mineral soil groups
19	Drainage
20	Erosion
49	Existing transportation

[a]Exclusionary criterion.

[4]"Windshield surveys" were conducted by Siting Commission staff from their vehicles; staff did not enter the sites during these surveys.

collection of the GIS data, to observe current land use, and to look for obvious exclusionary features.

The sites were evaluated first for characteristics such as wetlands, soils, geologic complexity, erosion, mineral resources, and proximate incompatible activities. Some of these criteria were duplicative of criteria used in the Map Assessments step (Table 6.2). Aerial photos were used to augment field observations (e.g., of heavily wooded or otherwise inaccessible areas). Supplemental technical information provided by local government units also was used in some cases (see discussion later in this chapter). A team of geological and mining experts performed an additional survey at selected sites being considered for the drift mine disposal method. Local geologists (not Siting Commission staff) also made a series of selected site visits. The details of this screening step (e.g., what combinations or levels of unfavorable conditions would exclude a site) were not described fully in the ROPSI. Based on this procedure, the Siting Commission excluded 16 of the 55 sites.

The remaining sites were then rescreened using the following criteria: drainage, presence of hydric soils, ground water discharge within the site, presence of nonresident populations, and slopes that were too steep for aboveground or belowground methods and too shallow for drift mines. This evaluation eliminated an additional 20 sites. Again, the basis for making site selections, or for excluding sites, was not discussed fully in the ROPSI.

RESCREENING USING ALL CRITERIA

Field screening reduced the number of potential sites to 19. These were located in the Allegany, Chenango North, Cortland, Montgomery, and Washington candidate areas (Figure 5.4). The Siting Commission rescreened these sites by applying the entire set of exclusionary and preference criteria identified in the Siting Plan (Appendix I).[5] The ROPSI summarizes the results of this screening as follows (p. 9-1):

---

[5]The Siting Commission never applied criterion 9, a preference criterion relating to existing mines, because existing mines were excluded before this criterion had been used.

> As a result of the comparative assessments described in the previous sections, staff concluded that many of the 19 sites appeared to be technically excellent and had a high potential for ultimately being found suitable for certification and licensing as a low-level radioactive waste disposal facility. Nevertheless, in moving forward into the next phase of the siting process, staff also recognized that schedule and cost constraints dictate that attention must be focused on a limited number of sites that appear most suitable. For these reasons, five sites with greatest potential, in staff's judgment, are recommended; the others would be set aside along [with] the sites previously considered.

Thus, 14 of the 19 sites were set aside by a staff decision, leaving 5 potential sites for further study (Table 6.3; Figure 6.1). Although in the ROPSI the Siting Commission describes some of the factors considered in making this decision and provides some explanation for why 14 areas were set aside, the details of the screening and the basis for the selection of the 5 potential sites are not described.

SITE CHARACTERIZATION

During the last step of phase 1 of the screening process (Figure 3.1), additional field studies were to be undertaken of the five potential sites to further assess their suitability. These studies were to have included limited drilling and surface assessments. As noted in Chapter 3, however, the screening process was halted by the governor before these investigations could be initiated.

Once these field studies had been completed, at least two sites were to be identified for more detailed characterization. Full site characterization was to last at least a year to allow technical experts to observe changes at the sites over an annual cycle. Studies were to address geology, ground and surface water conditions, weather patterns, plants and animals, population, land use, archeological and cultural resources,

TABLE 6.3 Final Potential Sites (after ROPSI, Table S-1)

Site Name	County	Town(s)	Size (Acres)
Taylor Central	Cortland	Taylor	473
Taylor North	Cortland	Taylor	686
West Almond	Allegany	West Almond	918
Caneadea	Allegany	Caneadea, Allen	1,039
Allen	Allegany	Allen	780
Total			3,896

transportation, and other natural and human features of the sites. Information gathered and analyzed during site characterization was to provide the necessary data for the Siting Commission to recommend one or more sites for certification.

VOLUNTEER AND OFFERED SITES

In its Siting Plan the Siting Commission recognized that landowners might "volunteer" land for an LLRW facility. The Siting Plan indicated (pp. 2-5 and 2-6) that

> such volunteers will be subject to the same technical and regulatory requirements applicable to other areas and sites. . . . An assessment will be performed first against statewide exclusionary criteria to determine whether the recommended site or area merits further consideration. If these criteria are satisfied, the volunteer will be carried through subsequent stages of the screening process. Unless exclusionary conditions rule out a volunteer site supported by a community, every effort will be made to keep it under consideration until the point where it can be compared with other potentially suitable sites in terms of technical suitability and other decision factors.

> Recommended sites without community support must be obviously superior compared to other potential sites in order to continue evaluations beyond the initial steps.

During the GIS Screening, five parcels of land were offered by private landowners.[6] None of the landowners offered evidence that these sites had support from their surrounding communities. In its consideration of these properties, the Siting Commission made several amendments to the procedures of the Siting Plan in a resolution passed at its January 1989 meeting. The resolution distinguished between *offered sites,* which were offered by landowners without community support, and *volunteer sites,* which were offered by landowners with the support of the surrounding communities. The January 1989 resolution established new procedures for determining the suitability of both types of sites.

Unlike the original Siting Plan, however, the language of the resolution was vague about the particular screening procedures that would be applied to the volunteer and offered sites. On this point, the newly stated procedure was only to

> perform initial in-house review against technical criteria to determine that the parcel meets the cutoff level in use at the current selection stage. The parcel must be at least as good as the sites being considered at that stage.

Four of the offered sites were added to the list of 51 sites being evaluated during the Field Surveys step. The offered site in Cortland County (Taylor North) was included among the five final potential sites (Table 6.3; Figure 6.1).

CONTRIBUTED INFORMATION

Following the conclusion of the CAI screening (Chapter 5), the Siting Commission requested information from local governments on each of the 10 candidate areas identified, including data on the following:

---

[6]According to the ROPSI, four of the sites are located in Allegany, Clinton, Cortland, and Franklin counties. The location of the fifth site is not given.

population, zoning, local climate, well records, and locations of recreational facilities, ecologically sensitive areas, historic zones, large private institutions, schools, hospitals, tourist attractions, inactive waste disposal sites, and floodplains. The contributed information was reviewed to determine whether it affected the sites under consideration in PSI screening. Some of this information was used during the windshield surveys and some during the Rescreening step to verify previous Siting Commission conclusions. Most of the information, however, was set aside for site characterization activities and preparation of the draft environmental impact statement.

In their responses to this request, most of the counties criticized the siting process and argued that no potential sites could be found in their candidate areas. The counties believed that the contributed data could be used to exclude their areas from consideration, just as much of the state had been excluded in previous screening steps. The Siting Commission, however, decided that most of the data were too detailed to be used at this stage and preferred to defer application of the information until characterization, after potential sites had been chosen.

A description of the information contributed by local governments is summarized in the ROPSI. On the basis of information received on the sites in Allegany County, two surface water bodies and watersheds were excluded, and the location of a surface water intake was corrected in the Siting Commission database. Extensive information was received from Cortland County, but the Siting Commission decided not to utilize most of it for PSI screening. Cortland County argued that the 1980 census was outdated and that the 1986 census should have been used. The Siting Commission defended the use of the 1980 census data, noting that they were available for all candidate areas.

Washington County submitted information on regional geology and cultural and historic resources. The Siting Commission used the regional geology in its windshield surveys. Cayuga County conducted a GIS analysis similar to the Siting Commission's exercise. That analysis and other contributed information were reviewed by the Siting Commission to verify its GIS database. Oswego County also conducted a GIS analysis and produced a composite map showing areas that should be excluded. These results corresponded closely with the Siting Commission's analysis. Chenango County provided general background data on the candidate areas. The county argued that the lack of information on the local ground water hydrology was a reason to exclude its candidate area. The Siting Commission responded that a thorough study of ground water would be conducted during the site characterization step if a potential site was identified.

SENSITIVITY ANALYSES

During GIS Screening, the Siting Commission performed a rudimentary sensitivity analysis to test the importance of three exclusionary criteria (44—mineral soil groups; 17—wetlands exclusion; and 6—reforestation areas). The purpose of the analysis was to determine the sensitivity of these criteria in siting decisions, and particularly to determine whether more potential sites would be identified if these criteria were relaxed. As a result of this analysis, the Siting Commission decided that the exclusion of all soils in mineral soil groups 1 through 4 was too restrictive. This criterion was redefined, based on guidance from the state's Department of Environmental Conservation (DEC) and the Department of Agriculture and Markets, to exclude only farms with at least $10,000 of annual production. The Siting Commission also found that criteria 17 and 6 excluded a significant number of sites, but because these exclusionary criteria were based on state laws or regulations, the commission decided that it had no flexibility to change them.

PUBLIC PARTICIPATION

During PSI, the Siting Commission's work focused increasingly on the selection of a small number of sites for an LLRW disposal facility. Correspondingly, its relations with citizens of communities living around these sites became more contentious. Over the summer of 1989, the Siting Commission held meetings with task force groups that had been formed by local governments to study disposal issues. These groups were supposed to obtain comments from the public and make recommendations on the conditions under which a proposed facility would be acceptable to their communities.

In September 1989 the Siting Commission released the ROPSI, which included its recommendations for the five potential sites. The commission had sent advance copies of the ROPSI to selected communities in Allegany and Cortland counties, and it invited officials to meet to discuss the report. This effort had an unanticipated effect—news of the site selections was leaked to the public, sparking protests in Allegany and Cortland counties before the official announcement.

The Siting Commission made an effort to reach out to affected communities in order to obtain public input and strengthen public awareness of all aspects of its work. These efforts were unsuccessful, however. Many of the Siting Commission's attempts to schedule meetings in affected communities were rejected. Invitations to nominate county representatives for the Advisory Committee (see Chapter 3) were also rejected. The Siting Commission encountered a high level of opposition at a public meeting held in Cortland County in November 1989. Also in November, the Cortland County Legislature passed a resolution calling for the Siting Commission to cease consideration of the two sites in the county.

A public participation status report was issued by the Siting Commission on January 19, 1990. It summarized activities up to that time and outlined future outreach efforts. The Siting Commission recognized that it could not overcome all opposition to siting, so it decided to concentrate its outreach efforts on less vocal segments of the affected communities. The commission planned to increase its presence in these communities, disseminate information of local concern, improve responses to public questions and comments, and coordinate technical and field activities so that accurate information would be provided.

The Siting Commission's difficulty in responding to the large volume of public questions and comments added to criticisms of the siting process. Members of the public complained that draft documents were not available for review or comment, that comment periods were too short, that final reports were not available, and that there were long delays in responding to requests for information. By the latter part of 1989, Siting Commission staff were apparently unable to handle the volume of correspondence, and a backlog developed. Between September 1989 and January 1990, more than 3,500 pieces of correspondence were handled by the Siting Commission; 90 individual letters were answered, more than 200 letters transmitting reports and program updates were sent to elected officials and task forces, and 120 letters were mailed to civic and environmental organizations.

The formal public comment period on the ROPSI had generated more than 510 pieces of written correspondence, including many multipage reports and documents. The Siting Commission estimated that more than 2,000 questions and comments would have to be addressed in its comment/response document. Some of the letters, especially those from Cortland County, required detailed answers.

In general, the Siting Commission was not welcomed in the affected communities. In December 1989 the commission's efforts to perform field investigations were blocked by citizens opposed to a disposal facility. The attempt to conduct inspections was ended in January 1990 because of public opposition. The Siting Commission's January 1990 *Public Participation Status Report* indicates that at least four attempts were made to meet with Cortland County officials in November and December 1989. One meeting was canceled by Cortland County, and an invitation to meet with the Siting Commission for a site walkover was declined. A meeting of Siting Commission staff and Cortland County officials and technical staff was finally held to discuss technical issues in February 1990. The videotape of the meeting reveals a chaotic scene at the morning session. Siting Commission staff attempted to make a presentation but were not allowed to use the overhead projector and were interrupted by protesters holding signs.

The Siting Commission opened a public information office in Cortland County and a temporary trailer office in Allegany County. There were daily public protests leading to numerous arrests. In April 1990 the governor closed these offices and ceased siting activities.

The citizens of Cortland and Allegany counties also sought legal relief in their dispute with the Siting Commission. Two suits were brought under freedom-of-information laws for GIS records, correspondence between the DEC and the Siting Commission, and a mine feasibility study prepared by the contractor, Roy F. Weston, Inc. Another suit alleged that the Siting Commission had violated the open-meetings law. Lawsuits challenged the Siting Commission's right to enter private property without a landowner's permission. Cortland and Allegany counties also joined with New York State in challenging the federal legislation that required states to take title to waste produced in their states after January 1, 1993 (see

Chapter 2). As noted in Chapter 2, the U.S. Supreme Court struck down that provision in 1992.

Citizens were also successful in changing New York's siting law. After suspending the activities of the Siting Commission, the governor introduced legislation in June 1990 to amend the 1986 State Act that established New York's siting process (Chapter 2). The act, which took effect in July, involved the following changes:

- The Advisory Committee was reformed into the Citizens Advisory Committee, with enhanced citizen membership and a greater role in facilitating public comment and review regarding the work of the Siting Commission.
- Public participation and review of the Siting Commission's work were expanded.
- An independent technical and scientific review panel was created to review the work of the Siting Commission.[7]
- Two new members were to be added to the Siting Commission: a social scientist and a person with environmental interests.
- A report to review the rationale for the exclusionary criteria was commissioned. This report, the *Excluded Areas Report*, was issued by the Siting Commission in August 1993.
- The act required that the disposal method be identified before the site selection work.

ANALYSIS AND DISCUSSION

As noted previously, all 17 exclusionary criteria and 43 preference criteria were applied during PSI screening. The committee identified problems with several of these criteria. The problems with some of the criteria applied during CAI screening were discussed in Chapter 5. The remainder are discussed below.

---

[7]As noted in Chapter 1, this legislation resulted in the formation of this National Research Council committee.

Criterion 1—Geologic Complexity

Prefer areas that contain uniformly distributed soils, sediments, and/or rock that are relatively undeformed and devoid of fractures, faults, and other discontinuities that may influence predictions of performance of the facility.

The intent of this criterion is to select sites with relatively uniform geology in order to simplify assessments of long-term performance. Such assessments usually involve modeling exercises to predict radionuclide transport times and pathways through the subsurface.

The Siting Commission made an implicit—and in the committee's opinion, a simplistic—assumption, namely, that uniform geology provides simple or predictable subsurface flow pathways. Flow complexity depends on the distribution of porosity and permeability, which is partly related to rock type and state of deformation and partly to fracture characteristics (spacing, length, interconnectivity, and orientation), particularly in bedrock.[8] In bedrock, much of the ground water in New York State is transmitted by fractures, and all previous deformation is unimportant unless it involved fracture generation. Very few rocks of this region have a porosity and permeability generated during the formation of the rocks; some (limestone, rock salt; see next section) have a porosity and permeability generated by solutions reacting with the rocks after their formation. In its response to the committee's questions (Appendix F), it is clear that the Siting Commission did not recognize that modeling capability depends on more than geologic and lithologic complexity, terms that are themselves vague and scale dependent. For a review of the current state of modeling and its applications, see NRC (1990) and Sullivan and Chehata (1995).

Another problem with this criterion is that it was difficult to apply without site-specific data. This criterion was scored on the basis of surrogate characteristics such as rock type and structural features such as

[8] *Porosity* is the percentage of the volume of a soil, rock, or sediment, that is made up of pore, or open, space. *Permeability* is the capacity of a substance to transmit fluids. It depends on the size, shape, and interconnectedness of pores.

folding and regional faults. In the committee's judgment, such superficial assessments, based on maps with limited resolution, were inappropriate at this stage of screening.

Criterion 3—Subsurface Dissolution

Prefer areas that do not demonstrate significant past or active subsurface dissolution.

Subsurface dissolution is a process that affects carbonate rocks (limestone and dolomite) and evaporites (halite, gypsum, and anhydrite). This criterion was designated as a preference criterion, although DEC regulations indicate clearly that it should have been an exclusionary condition. Specifically, the DEC regulations state:

> The disposal site must not be located in an area where subsurface hydrogeologic processes, such as dissolution, occur with such frequency and extent [as] to adversely affect the ability of the underground mined repository to meet the performance objectives of Subpart C of this Part. (6 NYCRR 382.23(c)(2))

An exclusionary criterion is warranted because of the significant performance implications of dissolution. Past or present subsurface solution can produce large subsurface voids, leading to subsidence or collapse. In the committee's opinion, no LLRW disposal site should be located where there is any indication of dissolution, regardless of disposal methodology. Geologic maps and cross sections of New York show the locations of these potentially soluble rocks both on and below the surface, and the presence of dissolution features can be ascertained through field studies.

Criterion 20—Erosion

Prefer sites exhibiting no existing or potential erosional characteristics that could adversely affect waste containment.

This is another example of a preference criterion for which there is a clear regulatory requirement addressed to an important performance specification. Specifically, DEC regulations state:

> The site must not be located in areas where the surface runoff could expose, erode or inundate the disposal units. (6 NYCRR 382.22(a)(4))

In the judgment of the committee, LLRW disposal facilities should not be located in areas with a high potential for surface erosion. Thus, this should have been an exclusionary criterion.

Criterion 44—Mineral Soil Groups

> *Exclude all lands in mineral soil groups 1–4, as designated by the New York State Land Classification System, that are in active agricultural production.*

This criterion is based on a system for quantifying the agricultural productivity of soils.[9] The Siting Commission implemented this criterion by excluding sites with more than 5 acres of land in active agricultural use with mineral soil groups 1 through 4. Active agricultural use was defined as use resulting in the production of more than $10,000 of income per farm per year from farming as reflected in agricultural tax exemption records.

For the GIS Screening analysis, the center points of all lands in mineral soil groups 1 through 4 were entered into the GIS using a cell size of 40 acres. A cell was considered to contain soils in these groups if the center point (centroid) of an area with these soils fell within the cell. The centroids and acreages of tax-exempt farm properties for 1988 were obtained from the New York State Department of Equalization and Assessment. Squares representing the acreages were entered into the GIS around the center points of the farms. If the square representing a farm fell into a cell containing mineral soil groups 1 through 4, that cell was eliminated.

---

[9]The classification is based on the productivity and the total digestible nutrients of the soil. The index values range from 1 to 10, where 1 is the highest grade.

The problem with this methodology is that it did not exclude all of the cells containing soils in these groups. If the centroid of an area with soils in groups 1 through 4 was near the edge of a 40-acre cell, for example, the adjacent nonexcluded cell could have contained significant acreage with the soil groups that should have been excluded.

Indeed, Criterion 44 was reevaluated by the Siting Commission during its Rescreening of the 19 sites. The precise polygonal boundaries of all soils within groups 1 through 4 were digitized from Soil Conservation Service (now Natural Resources Conservation Service) county soil surveys and matched to the GIS cells. The results showed that only 2 of the 19 sites had no soils in mineral groups 1 through 4. Three of the sites contained soils in groups 1 and 2, the most productive agricultural land, and the remaining sites contained soils in groups 3 and 4. Thus, in applying this criterion as it did, the Siting Commission deviated from its previous practices of removing cells that contained any part of an exclusionary feature.

Surface and Subsurface Drainage

Regulations for surface and subsurface drainage around LLRW facilities are provided by the U.S. Nuclear Regulatory Commission (USNRC) in 10 CFR 61.50(a)(5) and 61.51(a)(6) as follows, respectively:

> The disposal site must be generally well drained and free of areas of flooding or frequent ponding. Waste disposal shall not take place in a 100-year flood plain, coastal high-hazard area or wetland, as defined in Executive Order 11988 Floodplain Management Guidelines.
>
> The disposal site must be designed to minimize to the extent practicable the contact of water with waste during storage, the contact of standing water with waste during disposal, and the contact of percolating or standing water after disposal.

The corresponding DEC regulations are given in 6 NYCRR 382.

The Siting Commission's strategy for meeting these regulations was to site the facility in an area with low annual precipitation (Criterion 23), good surface drainage (Criterion 19), and impermeable but sorptive[10] stratigraphic units (Criterion 14). This approach is effective for meeting two of the requirements implicit in the regulations: (1) minimizing the percolation of surface and subsurface water into the facility and (2) minimizing radionuclide migration out of the facility and into the environment should water come in contact with the waste after disposal. This strategy may be ineffective, however, for meeting the third requirement of the regulations: minimizing the contact of standing water with waste after disposal.

It is important to recognize that an impermeable stratigraphic unit provides a barrier to water flow in all directions. This barrier keeps water out of a facility, but it also prevents any water that makes its way into the facility—for example, by the failure of the repository cap—from escaping into the subsurface. Thus, water can accumulate in a facility developed in an impermeable stratigraphic unit, absent active pumping.[11] If the repository is located near the surface, this water may eventually fill the facility and overflow into the surface environment.

A strategy for minimizing the contact of standing water with waste would be to develop the facility in more permeable stratigraphic units with proven sorptive capacity in order to promote the subsurface drainage of water entering the facility. To some, it may seem counterintuitive to develop an LLRW facility in a permeable stratigraphic unit through which radionuclide-laden water could "flush" into the water table, but, in fact, permeable and sorptive stratigraphic units, if present, may provide excellent performance as one component of a comprehensive system design. Such a design should also include a facility with an impermeable cap (natural or purpose-built) and good surface drainage to divert surface water from the facility. In addition, the

---

[10]*Sorption* refers to the process by which radionuclides or other contaminants are attached to the surfaces of host particles (e.g., clay particles) by local electrostatic charges or surface complexation reactions.

[11]This so-called "bathtub effect" has been observed in other LLRW facilities, most notably the West Valley, New York, and Maxey Flats, Kentucky, facilities (see Chapter 2).

facility could be placed at a site that was well above the water table to prevent subsurface water intrusion into the facility.

The committee concludes that the design of "drainage" criteria, particularly Criterion 14 was unnecessarily restrictive. Indeed, the application of this criterion during the PSI stage of screening may have eliminated some otherwise suitable areas of the state from consideration.

GIS Screening

The committee identified one problem with the GIS Screening step of the siting process: namely, that it employed an arbitrary cutoff score to select sites. As with CAI Initial Preference Screening, the committee questions the use of a cutoff score when a high score may correlate only weakly with adequate site performance (see Chapter 5 and the discussion of sensitivity analyses later in this chapter). Also, although the cutoff score was chosen to produce a manageable number of sites, it should be noted that the 3,900-point cutoff score chosen was substantially lower than the 4,400-point score used in the CAI Initial Preference Screening step.

Map Assessments

The committee had several concerns regarding the Qualitative Map Assessments step of screening. It should be noted that this step was not discussed in the Siting Plan but was added later in the process. The committee's primary concern with the Map Assessments step is that the screening methodology transformed preference criteria into exclusionary conditions. A single negative mark for one of the five preference criteria was sufficient to exclude a site. Moreover, the screening process selected sites with some limitations (i.e., sites with an "A" or "B" rating), yet there was no attempt to assess whether the limitations outweighed the positive attributes of the sites. An additional problem with this screening step is that some of the criteria were sensitive to the age of the data. This is of particular concern for data regarding proximity to incompatible activities and nonresident populations. Thus, these criteria may have been applied nonuniformly between areas.

Field Surveys

The committee has reviewed the plan of the surveys, the training of the staff performing the surveys, and the factors to be evaluated. All appear to be conceptually sound, but the committee did not obtain sufficient data to assess their application in screening.

The Siting Commission provided the committee only with examples of windshield survey form results for the four offered sites. The committee received little direct information regarding the procedures and evaluations for the other 51 sites, and no information was provided for the mine suitability evaluations.

These surveys were conducted over a two-week period. The ROPSI provides the windshield survey results only for those sites that were set aside, not for those that were kept. Thus, it is difficult for the committee to assess how the Siting Commission used this information to make its siting decisions.

The data collection forms for the windshield surveys (see Appendix K) indicate that a broad range of information was to be collected. This information would complement the data evaluated during the Map Assessments step. However, it is unclear to the committee how this information was actually used in the site selection. The ROPSI is also vague on this point; the report states (ROPSI, p. 5-1):

> The purpose of the limited site inspections was to identify those sites that continued to appear technically superior, as measured against the criteria in the Site Selection Plan, in light of the added information gained through visual inspection of site conditions.

The committee asked the Siting Commission to provide documentation regarding the criteria for the windshield surveys. Its response to this issue focused on the information that was collected, not on how it was used. The use of the windshield surveys during PSI screening was controversial because of county allegations that information was collected that should have been used to exclude two of the final potential sites (Taylor North and Taylor Central; Table 6.3; Figure 6.1). Specifically, the survey forms for these sites show that newly constructed incompatible structures were observed close to (and within) the site boundaries. Such information would have given the sites a "–" score and disqualified them during Qualitative Map Assessments screening. Indeed, similar information was used to exclude three other sites during the map screening step. Thus, the committee judges that the Siting Commission did not adhere to its stated goals (or to good scientific practice) in the windshield surveys, that is, to evaluate the sites "in light of additional information gained through visual inspection."

The committee also notes that the Siting Commission's use of information gathered from windshield surveys to confirm the overall favorability of the sites with high GIS scores may not have been clear to the public. This may explain why many members of the public believed that denying access to the sites would severely hamper the Siting Commission's inspections and that the windshield surveys would yield very little information with which to judge the sites. It was too late for the Siting Commission to justify its judgment that the 19 sites carried forward had superior conditions for siting based on observations made during the surveys because the public had very little confidence in the commission's decision-making process.

Rescreening Using All Criteria

Despite the considerable effort made by the Siting Commission to develop a technically sound screening and scoring process, it appears that the commission departed from its own plan at this final stage of screening. The committee believes that the reduction to five sites without detailed supporting documentation and technical justification was a significant flaw in the site selection process.[12] This decision may have contributed to the public perception that the process moved quickly to "volunteer" sites, bypassing the technical process defined in the Siting Plan.

---

[12]Notably, the 5 potential sites selected in this step of the screening process (Table 6.3) were not the highest-scoring sites. The 5 sites were ranked 1, 5, 6, 7, and 13 for aboveground or belowground disposal and 1, 2, and 4 for mine disposal, based on scoring using the 43 preference criteria. Two of the sites were not under consideration for mine disposal. While these choices may well have reflected the Siting Commission's recognition of the weaknesses of the scoring system that have been noted elsewhere in this report, the commission did not explain its underlying logic.

This is the first (and only) stage of the screening process in which all 43 preference criteria are used for screening, and it is in this stage that some of the technical deficiencies of the scoring system first become apparent. One area of particular concern to the committee is the scaling factors used in scoring the sites. Scaling factors for most of the preference criteria used in this step of screening were assigned values of 1, 3, or 5 according to the conditions at the site, but there were several exceptions. For example, the scaling factor for Criterion 52—transportation safety—took values of 2, 3, or 4; Criterion 50—multimode access—took scaling factor values of 3, 4, or 5. Several criteria (e.g., 19—drainage; 33—low population densities; and 34—highly populated places) took scaling values of 1, 2, 3, 4, or 5.

The design of scaling factors is not documented in any of the Siting Commission's reports, and the committee questions whether the scaling factors used in scoring are justified technically. Although the committee recognizes that scaling factors are subjective, it particularly has concerns about the range of scaling factors applied to each criterion. A scale factor value of 5, for example, produced five times the number of points as a scale factor value of 1. Such a difference can produce a large change in the overall score and can have a significant effect on the suitability of the site—changes that may not be justified in all cases.

Consider the effects of these scaling factor values on Criterion 23—annual precipitation—for example. A locality with annual precipitation of 50 inches or more would receive a scale factor value of 1, whereas a locality with 39 inches per year would receive a scale factor value of 5. There is no evidence to demonstrate that such a nonlinear sensitivity to rainfall values is warranted by related impacts of rainfall on a low-level waste disposal site.

Volunteer and Offered Sites

Because Taylor North, an offered site, was among the five potential sites, there was considerable scrutiny of the Siting Commission's activities regarding the screening of offered land. Indeed, an August 1992 report by the U.S. General Accounting Office on the selection process for the Taylor North site concluded that "the New York Low-Level Radioactive Waste Siting Commission did not comply with the letter of its procedures for an offered site. . ." (U.S. General Accounting Office, 1992, p. 18).

From its own analysis, the committee believes that the Siting Commission's decision to include the Taylor North site in the final list of five potential sites was inconsistent with the commission's requirement that an offered site must be "at least as good as" other sites. The Taylor North site should have been excluded based on Criterion 44 because it had more than 5 acres of land in agricultural production in soils of groups 1 through 4.[13] Second, the site did not meet the minimum score that was applied at this stage of screening (3,900 in five contiguous 40-acre cells). The Taylor North site contained only two cells with scores above the cutoff. Eleven other cells had scores ranging from 3,709 to 3,879. Although this cutoff score of 3,900 points was arguably arbitrary, it was the value applied to all other sites in the PSI process and should have been applied to the offered lands as well.

Contributed Information

The committee reviewed the circumstances surrounding the solicitation of contributed information from affected communities by the Siting Commission. The committee believes that the commission made a reasonable effort to obtain contributed information. Further, the committee believes that the commission made a sincere effort to review and use information supplied to it by local governments. Information that was neither used directly nor used to verify information obtained from other sources was deferred for use during independent evaluation of sites rather than for comparison of sites with each other.

In hindsight, however, the Siting Commission's decision to defer consideration of these data was not successful from a public relations point of view, primarily because it appeared to local governments that their offered information was being ignored. Although the Siting Commission did endeavor to utilize all of the information supplied by local governments, it did not effectively communicate to local governments and the public why it decided to defer using much of the information until later

---

[13]The Taylor North site had 171 acres of land in group 4, most of which was in active agriculture. The Siting Commission noted in the ROPSI (p. 9-2) that a variance would be required to develop this site.

stages. The Siting Commission believed this decision was based on its best judgment of information available at the time, given the circumstances (the need to identify potential sites quickly); however, it would have been very difficult for the commission to fully explain how it used the information and to assure local governments that it was not making an arbitrary decision. Thus, critics of the process viewed this as just another example of decisions made without sound scientific and technical bases.

A review by the committee of materials submitted by the counties revealed no inconsistencies with data in the databases utilized by the Siting Commission to apply exclusionary criteria in the GIS Screening step of PSI screening. Given the definitions of the exclusionary criteria and the resolution of the databases, it appears that the Siting Commission correctly identified all of the features that should have been excluded during screening. The committee emphasizes, however, that this is a qualified conclusion; the committee reviewed materials data only for the 10 candidate areas, which represent a small portion (approximately 2 percent; see Chapter 5) of the state. The committee makes no judgments regarding the use of data during previous exclusionary screenings. The committee also makes no judgments regarding the use of the data in scoring the preference criteria in PSI screening. Scoring was a very controversial issue, but some degree of interpretation was inherent in the scoring process.

Sensitivity Analyses

For the PSI phase of screening, the Siting Commission focused its sensitivity analysis largely on criteria that were constrained by law or regulation, where few modifications could be made. As described in Chapter 5, the purpose of sensitivity analyses is to identify and remedy weaknesses in the screening process. When measured against this objective, the analysis undertaken by the Siting Commission was of limited value. It is important to recognize that the commission had considerable flexibility in the interpretation and implementation of the preference criteria used in PSI screening. Thus, a detailed sensitivity analysis would have been useful for assessing the sensitivity of the siting decision to these criteria.

The committee performed its own sensitivity analysis using groups of preference criteria applied in PSI screening (see Tables 6.4 and 6.5). This analysis is similar in design to that presented in Chapter 5, except that the full set of preference criteria are used. The preference criteria used in the sensitivity analysis are divided into three groups:

1. Preference criteria based on *known quantities*—readily available data having relatively uniform quality across the state. Most of these criteria address nonperformance factors, that is, factors that are not critical in determining the capability of a site to function as an LLRW repository.
2. Preference criteria based on *uncertain quantities*—data of unknown or uneven accuracy or completeness. Most of these preference criteria address the likely performance of the site, as noted in Chapter 5 (see particularly footnote 7 in Chapter 5). In the scoring system used by the Siting Commission (Chapter 3), these criteria tend to give high scores in regions that lack data. Scoring can thus be viewed from two perspectives. If complete information is available, these criteria represent critical qualities for a potential disposal site, and sites should not be able to qualify without favorable scores on them. Alternatively, for the case of incomplete information, the committee's analysis illustrates the scores that would be given to poorly characterized sites.
3. Preference criteria based on *population quantities*—in this case, demographic information. These quantities can be readily measured, but they are sensitive to the details of averaging as noted in Chapter 5.

The results of two sensitivity analyses performed by the committee are presented in Tables 6.4 and 6.5. Table 6.4 shows the sensitivity analysis for the criteria employed in GIS Screening. As noted earlier in this chapter, GIS Screening used a cutoff score of 3,900 points. Table 6.5 utilizes the entire set of preference criteria employed at the final stage of PSI screening. The Siting Commission did not use a cutoff score for this screening step.

As shown in Table 6.4, the weights of the three categories of preference criteria—known, unknown, and population quantities—are

TABLE 6.4 GIS Screening Sensitivity Analysis (cutoff score = 3,900 points)

Category (Criterion Number)	Criterion Weight[a]	Scaling Factor for Scenario[a] #1	#2	#3	#4	#5	#6	#7	#8
Known Quantities									
Distance from mines (7)	20	1	1	5	1	1	5	1	5
Surface water features (13)	40	1	1	5	1	1	5	5	5
Distance from wetlands (18)	30	1	1	1	1	1	1	5	5
Annual precipitation (23)	20	1	1	1	1	1	1	1	5
Chronic severe weather (24)	15	1	1	1	1	1	1	1	5
Severe weather frequency (25)	20	1	1	1	1	1	1	1	5
Other radionuclide sources (30)	20	1	1	1	1	1	1	1	5
Distance from federal lands (37)	10	1	1	5	1	1	5	5	5
Distance from state lands (39)	10	1	1	5	1	1	5	5	5
Distance from Indian lands (42)	10	1	1	5	1	1	5	5	5
Transportation access (47)	10	1	1	1	1	1	1	1	5
Multimode access (50)	5	3	3	3	3	3	3	3	5
Proximity to waste generators (51)	20	1	1	1	1	1	1	1	5
Subtotal	230	240	240	600	240	240	600	640	1,150
Uncertain Quantities									
Geologic complexity (1)	45	1	1	1	5	5	5	5	5
Seismic hazards (2)	20	1	1	1	1	1	5	5	5
Subsurface dissolution (3)	35	1	1	1	5	5	5	5	1

Geologic units (5)	35	1	1	1	5	5	5	5	3
Mineral resource potential (8)	20	1	1	1	5	5	5	1	5
Distance from oil and gas fields (10)	20	1	1	1	5	5	5	1	5
Primary and principal aquifers (12)	55	1	1	1	5	5	5	5	1
Unconsolidated stratigraphic units (14)	40	1	1	1	5	5	5	5	1
Best usage of surface waters (22)	30	1	1	1	5	5	5	5	1
Ecology (29)	55	1	1	1	5	5	5	5	5
Distance from historical/cultural resources (59)	20	1	1	1	5	5	5	1	5
Subtotal	375	375	375	375	1,795	1,795	1,875	1,635	1,165
Population Quantities									
Low population densities (33)	45	1	5	5	1	5	1	5	5
Highly populated places (34)	45	1	5	5	1	5	1	5	5
Routes through incorporated places (53)	10	1	5	5	1	5	1	1	5
Subtotal	100	100	500	500	100	500	100	460	500
Total Score	705	715	1,115	1,475	2,135	2,535	2,575	2,735	2,815
Normalized Score[b]	1,001[c]	1,015	1,583	2,095	3,032	3,600	3,657	3,884	3,997

[a]See Chapter 3 for an explanation of these quantities.
[b]Normalized scores were determined by multiplying by a factor of 1.42. This factor represents the ratio of the maximum score using this subset of criteria (maximum score = 3,525 when siting factor = 5 for all criteria) to the maximum score using the complete set of preference criteria (maximum score = 5,000 when siting factor = 5 for all criteria), as documented in the ROPSI (p. 4-45). In CAI Initial Preference Screening, weights were renormalized to 1,000 points before scores were determined.
[c]This score was reported as 1,000 in the ROPSI.

TABLE 6.5 Potential Sites Rescreening Sensitivity Analysis

Category (Criterion Number)	Criterion Weight[a]	Scaling Factor for Scenario[a] #1	#2	#3	#4	#5	#6	#7	#8
Known Quantities									
Distance from mines (7)	20	1	1	5	1	1	5	5	5
Surface water features (13)	40	1	1	5	1	1	5	5	5
Distance from wetlands (18)	30	1	1	1	1	1	1	5	1
Drainage (19)	30	1	1	1	1	1	1	5	1
Flooding (21)	25	1	1	1	1	1	1	5	1
Annual precipitation (23)	20	1	1	1	1	1	1	5	1
Chronic severe weather (24)	15	1	1	1	1	1	1	5	1
Severe weather frequency (25)	20	1	1	1	1	1	1	5	1
Other radionuclide sources (30)	20	1	1	1	1	1	1	5	1
Distance from federal lands (37)	10	1	1	5	1	1	5	5	5
Distance from state lands (39)	10	1	1	5	1	1	5	5	5
Distance from Indian lands (42)	10	1	1	5	1	1	5	5	5
Government-owned lands (43)	20	1	1	5	1	1	5	5	5
Transportation access (47)	10	1	1	1	1	1	1	5	1
Existing transportation (49)	5	1	1	1	1	1	1	5	1
Multimode access (50)	5	3	3	3	3	3	3	5	3
Proximity to waste generators (51)	20	1	1	1	1	1	1	5	1
Labor force (54)	10	1	1	1	1	1	1	5	1
Housing stock (55)	10	1	1	1	1	1	1	5	1
Municipal services (56)	15	1	1	1	1	1	1	5	1
Viewshed (60)	15	1	1	1	1	1	1	5	1
Subtotal	360	370	370	810	370	370	810	1,800	810
Uncertain Quantities									
Geologic complexity (1)	45	1	1	1	5	5	5	1	5
Seismic hazards (2)	20	1	1	1	1	1	1	5	1
Subsurface dissolution (3)	35	1	1	1	5	5	5	1	5

Geologic units (5)	35	1	1	1	5	5	5	1	5
Mineral resource potential (8)	20	1	1	1	5	5	5	5	5
Distance from oil and gas fields (10)	20	1	1	1	5	5	5	5	5
Primary and principal aquifers (12)	55	1	1	1	5	5	5	1	5
Unconsolidated stratigraphic units (14)	40	1	1	1	5	5	5	1	5
Erosion (20)	35	1	1	1	5	5	5	1	5
Best usage of Surface waters (22)	30	1	1	1	5	5	5	1	5
Ecology (29)	55	1	1	1	5	5	5	1	5
Distance from historical/cultural resources (59)	20	1	1	1	5	5	5	1	5
Subtotal	410	410	410	410	1,970	1,970	1,970	650	1,970
Population Quantities									
PSD increment (27)	30	1	5	5	1	5	1	5	5
Proximity to incompatible activities (31)	20	1	5	5	1	5	1	5	5
Low population densities (33)	45	1	5	5	1	5	1	5	5
Highly populated places (34)	45	1	5	5	1	5	1	5	5
Nonresident populations (35)	25	1	5	5	1	5	1	5	5
Development/population growth (45)	20	1	5	5	1	5	1	5	5
Traffic congestion (48)	10	1	5	5	1	5	1	5	5
Transportation safety (52)	15	2	4	4	2	4	2	4	4
Routes through incorporated places (53)	10	1	5	5	1	5	1	5	5
Noise sensitivity (61)	10	1	5	5	1	5	1	5	5
Subtotal	230	245	1,135	1,135	245	1,135	245	1,135	1,135
Total Score	1,000	1,025	1,915	2,355	2,585	3,475	3,025	3,585	3,915

NOTE: PSD = Prevention of Significant Deterioration.

[a] See Chapter 3 for an explanation of these quantities.

230, 375, and 100, respectively. The relative differences in these weights are similar to those in the CAI Initial Preference Screening (see Table 5.5), which were 347, 533, and 120. Consequently, many of the conclusions of the sensitivity analysis for CAI Initial Preference Screening also apply here. Most notably, preference criteria scored using "uncertain quantities" continue to play a significant role in PSI screening, indicating that there may be a weak correlation between the final score and the suitability of a site for an LLRW disposal facility.

The use of a lower cutoff score for GIS Screening in PSI (3,900 points) than for Initial Preference Screening in CAI (4,400 points) allowed sites to exceed the cutoff with less favorable overall ratings. Most significantly, sites could receive less favorable or even unfavorable scores on important performance-related criteria and still exceed the cutoff. This situation is illustrated in Scenario 8 in Table 6.4. Many of the critical performance criteria are ranked as unfavorable [scaling factor (sf) = 1], yet the total score (3,997 points) still exceeds the cutoff. In contrast, these critical performance criteria are scaled as most favorable (sf = 5) in Scenarios 6 and 7, yet the total scores (3,884 and 3,657 points) fall below the cutoff because of unfavorable ratings on other criteria. The score in Scenario 7 can be boosted above the cutoff by an inconsequential change in a preference criterion such as annual precipitation.

GIS Screening had a modest bias toward rural areas. The DEC regulations also had a bias toward rural areas, but it is not clear that the Siting Commission fully recognized the additional degree of bias it introduced through the use of multiple criteria. The total weight of population-related criteria was 100 out of 750 points. The other criteria used in this step of screening correlated only weakly with population criteria.

Table 6.5 shows the sensitivity in scores to the full range of preference criteria that were used in the Rescreening step of PSI. The weights of the three categories of preference criteria (i.e., known, unknown, and population quantities) are 360, 410, and 230. Population and associated socioeconomic criteria take on a much greater significance in this screening step; that is, they account for more of the total weight. The significance of these criteria is even greater than that indicated by the increase in weight, however, because many of these criteria are correlated, as explained below.

In general, the population criteria give high scores to rural areas with low populations, with the possible exception of the Criterion 35, concerning the presence of nonresident populations. For this reason, the committee concludes that scoring at this stage of screening is biased toward rural areas. It should again be noted that part of this bias is derived from a bias toward rural areas inherent in DEC regulations, but that it is not clear whether the Siting Commission fully appreciated the degree of bias that was present due to the combined effects of the full set of factors.

Preference criteria scored with "known quantities" are linked strongly to population. These criteria include existing transportation (Criterion 49), labor force (54), housing stock (55), and municipal services (56). In general, the scores for these criteria are inversely proportional to those for the population quantities: highly populated areas will tend to score highly on these factors. There are also a number of transportation criteria that are highly correlated. Thus, they will receive similar scores because they largely describe the same siting factors. These include transportation access (47), existing transportation (49), multimode access (50), traffic congestion (48), transportation safety (52), and routes through incorporated places (53). Together, these 6 criteria have a weighting of 55 points, comparable to or greater than the weights of many of the individual performance-related criteria.

Rescreening with the full set of preference criteria addressed a large number of socioeconomic issues. Socioeconomic criteria included Criteria 27, 31, 33, 34, 35, 45, 48, 52, 53, 54, 55, 56, 59, 60, and 61 (see Table I.2). These criteria, which addressed issues such as the available labor force, access to highways, and population density, account for 30 percent of the total weight (300 out of 1,000). The committee recognizes that socioeconomic issues are critical concerns in siting an LLRW disposal facility, but there are two problems with the way in which these issues were handled by the Siting Commission. First, the specific criteria applied in screening mainly addressed demands for local facilities and services when large numbers of workers move into a rural area—which would have been quite modest for any of the facilities being envisioned even in the most rural areas of New York State—as opposed to focusing on the specific socioeconomic concerns that had been identified as most important by the scientific literature of the day. The more relevant concerns, as well known even by the early 1980s, are those having to do with equity, local control, and susceptibility to what were called "special" impacts (see Chapter 7). Second, a site could receive unfavorable ratings for critical performance factors and yet still be selected if it scored highly on socioeconomic and other factors that were not so directly related to facility performance, as in the case of the rainfall example noted earlier.

Although a cutoff score was not used at this stage, the five selected potential sites had scores in the range of 3,585 to 4,125 points. As shown by Scenario 7 in Table 6.5, it would be possible to attain a score of 3,585 and still receive unfavorable scores on most of the performance-related criteria. The analysis in Table 6.5 also shows that a rural site with little baseline information could also score above this cutoff. This is illustrated by Scenario 8, which received favorable ratings (sf = 5) on all but two of the uncertain and population quantities and was located away from ground water discharge zones and federal, state, and Indian lands.

In the committee's judgment, site selection should be optimized for both performance-related criteria and socioeconomic criteria, but to the extent feasible and reasonable, the performance-related criteria should be applied first: that is, screening based on performance criteria should be carried out before screening based on socioeconomic or other criteria not directly related to facility performance. The committee recognizes that such a division will often prove more challenging than is first apparent because even the most precise screening using "performance-related" criteria involve an inescapable element of human judgment, particularly when it is necessary to consider those criteria in combination with one another. A 1981 assessment (reprinted in U.S. Office of Technology Assessment, 1985, p. 218) recognized that

> judgments about acceptability [of nuclear waste facilities] are fundamentally matters of preference. Scientific and technical findings can inform those judgments by clarifying what the levels of safety associated with particular system design or repository siting decisions are likely to be. Even if those findings should be consensually accepted as being empirically accurate (no small task in itself), it still remains for the individual or society as a whole to determine, based on a set of values, whether those levels of safety are satisfactory or not.

Although the Siting Commission combined performance-related and non-performance-related criteria in ways that ultimately proved to be problematic, the problem is not simply that socioeconomic criteria were explicitly included in the calculations. Rather, most of the attention to socioeconomic criteria was focused on matters of relatively minor concern. In addition, the Siting Commission did not do the kinds of sensitivity analyses that would have identified the degree to which its actual choice of sites was shaped by factors having little to do with the probable safety or performance of the site.

SUMMARY

The purpose of the PSI process was to screen the 10 candidate areas in 4 discrete steps in order to identify a limited number of potential sites for an LLRW disposal facility.

The committee analyzed the exclusionary and preference criteria used in PSI, and it also analyzed the PSI screening process itself. The committee's findings are summarized in the following sections.

Screening Criteria

All 17 exclusionary and 43 preference criteria were applied during PSI screening. The committee identified problems with several of these criteria, some of which are discussed in Chapter 5. The remainder are summarized below.

- The assumption underlying the design of *Criterion 1—geologic complexity*—namely, that uniform geology provides simple or predictable subsurface flow pathways, is overly simplistic. The application of this criterion at this stage of screening was inappropriate without more detailed, site-specific data.
- *Criterion 3—subsurface dissolution*—should have been applied as an exclusionary condition because of the significant performance implications of dissolution. Subsurface dissolution can produce large subsurface voids, leading to subsidence or collapse.
- *Criterion 20—erosion*—should have been applied as an exclusionary condition. In the judgment of the committee, LLRW disposal facility should not be located in areas with a high potential for surface erosion.
- The methodology used to apply *Criterion 44—mineral soil groups*—during screening was inappropriate, because it failed to exclude all areas under active agriculture containing these soils. At this point in the process, the Siting Commission deviated from its previous practices of removing cells that contained any part of an exclusionary feature.
- The design of drainage criteria, particularly *Criterion 14—unconsolidated stratigraphic units*—was unnecessarily restrictive and may have eliminated many otherwise suitable areas of the state from consideration.

Screening

PSI screening comprised four discrete steps: (1) GIS Screening, (2) Qualitative Map Assessments, (3) Field Surveys, and (4) Rescreening Using All Criteria. The committee's findings with respect to these screening steps are summarized below.

• *GIS Screening.* As with CAI Initial Preference Screening, in GIS Screening there may be only a weak correlation between scoring and the likely performance of the site.

• *Qualitative Map Assessments.* The Qualitative Map Assessments step was not discussed in the Siting Plan but was added at a later time by the Siting Commission. The primary technical problem with this screening step is that it treated preference criteria as exclusionary conditions in that a single unfavorable mark on any one of five preference criteria was sufficient to eliminate a site from consideration. Additionally, there appeared to be little attempt by the Siting Commission in selecting sites for further screening to assess whether each site's limitations outweighed its positive attributes, as determined by using the preference criteria.

• *Field Surveys.* The plan, staff training, and criteria applied in the Field Surveys step appear to the committee to be conceptually sound. There was not enough information provided to the committee, however, for it to determine how the information collected in these surveys was actually used in site selection. There are some indications that data from these surveys were not used appropriately in at least some cases. For example, field survey data collected for the Taylor sites (which were among the five final sites; Table 6.3; Figure 6.1) indicate that they should have been excluded because of the presence of incompatible structures.

• *Rescreening Using All Criteria.* The selection of five potential sites was based on an undocumented staff decision, a decision that was not adequately described in the ROPSI. The design of scaling factors is not documented adequately in any of the Siting Commission's reports.

Although the committee recognizes that, at best, scaling factors are subjective, it questions whether many of the scaling factors are based on reasonable technical considerations.

• *Volunteer and Offered Sites.* The Siting Commission's decision to include the Taylor North site in the final list of potential sites (Table 6.3; Figure 6.1) is inconsistent with the commission's requirement that offered sites must be at least as good as other sites. The Taylor North site did not meet the minimum cutoff score applied to other sites at the same stage of screening. Furthermore, the site contained soils in mineral soil groups 1 through 4 and was under active agriculture (see footnote 13), which should have disqualified it from further consideration.

• *Contributed Information.* In the committee's judgment, the Siting Commission made a sincere effort to obtain, review, and use contributed information supplied to it by local governments, but the commission did not effectively communicate why it decided to defer using much of the information until later stages. The committee compared some of this contributed information with data in Siting Commission databases and found no inconsistencies.

• *Sensitivity Analyses.* The sensitivity analysis undertaken by the Siting Commission was of limited value because it was directed largely at criteria that were specified by law or regulation. The Siting Commission had considerable flexibility in the implementation of preference criteria at this stage of screening. Consequently, it should have performed a detailed sensitivity analysis to address the effects of its scoring system on siting decisions.

The committee performed its own sensitivity analysis for two steps of PSI screening (GIS Screening and Rescreening) in order to examine the effects of the weighting and scaling factors on scoring. Its conclusions are summarized below:

1. Preference criteria scored using uncertain quantities (i.e., data of unknown or uneven accuracy and completeness) had a significant effect on scoring during both GIS Screening and Rescreening.

Consequently, there may be a weak correlation between the final score for a site and its suitability as an LLRW repository.

2. As with CAI Initial Preference Screening, sites could receive unfavorable scores on performance-oriented criteria and still exceed the cutoff if they scored well on socioeconomic criteria. Sites with scores near the cutoff could be pushed across the cutoff threshold by inconsequential changes in certain criteria (see also point 4 below).

3. The GIS Screening and the Rescreening steps had a bias toward rural areas, which is partly reflective of a bias in the DEC regulations.

4. In the committee's judgment, scoring for the Rescreening step inappropriately combined performance and socioeconomic criteria. Thus, a site could receive less favorable or unfavorable ratings for important performance criteria yet still score above the cutoff if it received favorable scores on a large number of socioeconomic factors. The committee believes that site selection should have been based on a sequential screening using both performance-related and socioeconomic criteria after the exclusion of lands specified by regulations. The next screening step would apply performance criteria that are related most strongly to licensing and site performance, using a cutoff score selected to meet the desired area goal. Subsequent screening steps would apply those socioeconomic factors that are less strongly related to site performance in some logical and defensible order.

7
Discussion and Conclusions

As noted in Chapter 1 of this report, the New York siting process was halted by the state legislature in 1995 after an eight-year, $55.2 million effort, which was marked in its end stages by intense public scrutiny and criticism. The committee's purpose in this chapter is to address the causes of this failure, both internal to the Siting Commission and externally imposed. The committee also attempts to identify the lessons that might be drawn from the New York experience for the benefit of decision makers involved in future efforts to site nuclear facilities, of which those involving nuclear waste may be a particularly useful example.

EXTERNAL CAUSES

In the committee's judgment, a significant cause of failure of the siting process was the unrealistic schedules imposed on the Siting Commission by law and regulation. The federal and state legislation that created the framework for the Siting Commission's work (Chapter 2) contained strict time schedules with severe penalties for noncompliance. The Siting Commission was required to develop and carry out a process for identifying and characterizing a potentially certifiable site or sites for a low-level radioactive waste (LLRW) disposal facility within 19 months. The list of specific tasks for the Siting Commission was extensive (Chapter 2) and included site screening, submission of plans for site characterization studies to the Department of Environmental Conservation (DEC) for review, and site characterization and monitoring studies. The Siting Commission was also required to consult with DEC about the application for certification; submit for DEC's approval its proposed scope for a draft environmental impact statement; hold public meetings on the scope of the draft environmental impact statement; and submit copies of the application to DEC, the New York State Energy Research and Development Authority, the legislature, and the governor.

In the committee's opinion, the schedule imposed on the Siting Commission was unrealistic given the complexity of the project. In its review of the Siting Commission's work, the committee saw no indication that the commission fully realized the complexity of its task and the inherent conflict between its ambitious plan of work and the externally imposed time constraints. Perhaps because of the inexperience of its staff in siting activities, the commission's strict adherence to externally imposed schedules may have prevented it from undertaking the kind of careful planning and public outreach necessary to bring its work to a successful conclusion.

Although one could reasonably argue that the Siting Commission did what the regulations required it to do, in hindsight the Siting Commission would have been far better off had it made an effort early on to assess realistically whether its plan of work was achievable in light of others' experiences in siting such facilities. Such an assessment would probably have indicated that the New York effort was unlikely to succeed within the narrow time constraints allowed by law and regulation, and it may have prompted the Siting Commission to adopt a more deliberative siting approach following the examples of other states and compacts.

The difficulty of the Siting Commission's task was also increased by the intensity of public opposition. Some members of the public refused to play any role in the siting process. There was also an element of opposition that clearly and vocally stated that it would not accept any site, regardless of the technical justification.[1]

One of the key lessons to emerge from the accumulating experience with developing radioactive waste disposal sites has to do with what Freudenburg and Gramling (1994, p. 91) call "the spiral of stereotypes" which often occurs when persons on different sides of an issue stop talking to one another but not *about* one another (see also Coleman, 1957). The process of polarization had become sufficiently advanced in New York State by the time this committee was established that committee members experienced a good deal of pressure to take sides and to agree with the partisans on one side of the issue that the partisans on the other side deserved censure or blame.

---

[1]This sentiment was voiced by a number of members of the public at the open sessions held by this committee in New York State.

Research (see, for example, Galanter, 1974; Freudenburg and Pastor, 1992; Kunreuther et al., 1993) has shown that the initiators of projects have a great deal of influence over the degree to which the process becomes polarized or, alternatively, is characterized by relatively respectful, two-way communication. Proponents of future nuclear facilities will be well advised to avoid the adversarial escalation characteristic of the spiral of stereotypes and actively seek to develop better mutual understanding with facility opponents. Opponents of proposed facilities may not choose to cooperate with that approach, and even a siting process built on mutual respect and understanding cannot be guaranteed to succeed. However, evidence of the past 20 years indicates that approaches that contribute to a downward spiral in relationships are virtually guaranteed to fail.

INTERNAL CAUSES

As noted in this review, the Siting Commission played an important, although inadvertent, role in its own demise, owing to various deficiencies in the siting process and in its interactions with affected communities. In the committee's judgment, a primary reason for the failure of the siting effort was the siting process itself. The committee's review of the siting process (Chapters 3 to 6) revealed three significant technical flaws: (1) the goals of the process were ambiguous, (2) some aspects of the screening process were not well documented or technically defensible, and (3) strategic planning and quality assurance programs that could have identified and remedied these deficiencies were not put into place early enough in the process. These elements, combined with the unrealistic time frame, led to a strongly negative reaction from potential host communities that resulted in termination of the siting process.

Ambiguity of Goals

As noted in Chapter 3, the Siting Plan suggested that the goal of the screening process was "designed to identify sites that can satisfy the regulatory requirements and to demonstrate that *no obviously superior alternatives can be readily identified*" (Siting Plan, p. S-4, emphasis added). There may be many sites in New York that would meet the regulatory requirements to be certified and licensed. However, the demonstration that a site has no readily identifiable, superior alternatives is difficult at best. Many factors are involved in the suitability assessment, and no one site can be expected to be superior in all respects.

Part of this ambiguity may be due in part to perceptual differences among involved parties in what constitutes a site with no readily identifiable superior alternatives. To the professionally trained members of the Siting Commission, staff, and contractors, such a site could reasonably be perceived as one that meets the technical requirements for licensing and that is located in a community that will accept an LLRW disposal facility. Judging from the responses to the selection of candidate areas (see discussion in Chapter 5), however, many of the affected communities appeared to perceive such a site as "best" in some objective sense. This perception may have been reinforced by the use of a highly technical, quantitative screening process that sought to eliminate from consideration all but a handful of sites in the state. Whenever such a process is used, a decision to choose any site except the one with the "best score" is likely to create a need to justify or explain the choice.

Although the screening process may have led to the identification of licensable sites, the process itself was sufficiently complicated, with multiple screening steps involving different combinations of criteria and weighting factors, that it would have been very difficult for the Siting Commission to demonstrate to a skeptical public that superior sites did not, in fact, exist. The problems with the process identified in the latter stages of screening (e.g., the handling of offered sites; see Chapter 6) no doubt reinforced the perceptions of some that the process was not suited for identifying such sites.

Technical Problems with Screening Process

The committee identifies several technical problems with screening in Chapters 5 and 6 of this report. These include the following:

• Some of the screening criteria were flawed in their design or application.

• Data used to evaluate some of the performance-related criteria were of unknown quality and completeness. In some cases, the quality of these data was highly variable across the state. Thus, there may be a weak correlation between scores on these criteria and the likely performance of the site as an LLRW disposal facility. Moreover, the scoring system used by the Siting Commission favored the selection of areas with limited data for these criteria.

• Some of the criteria used in screening were not useful for discriminating the desired characteristics of the site or were combined in inappropriate ways. For example, performance and socioeconomic criteria were combined inappropriately during Candidate Area Identification (CAI) and Potential Sites Identification (PSI) screening (see discussions of sensitivity analyses in Chapters 5 and 6). Thus, a site could receive low scores on performance criteria, yet still exceed the cutoff if it scored highly on socioeconomic criteria.

• There was insufficient analysis to determine the sensitivity of the scoring system to coupling effects among screening criteria, the order in which criteria were applied, or small changes in scaling and weighting factors. The Siting Commission did carry out limited sensitivity analyses, but they are poorly documented, and the analyses addressed only a small fraction of screening criteria.

• The Siting Commission did not always follow its own procedures as defined in the Siting Plan. For example, the commission did not follow the Siting Plan in selecting five potential sites in PSI screening (Chapter 6) or in evaluating offered sites.

• The Siting Commission did not document its decision-making processes adequately, particularly regarding the identification of potential sites.

Strategic Planning and Quality Assurance

Many of the technical problems identified in this report could have been addressed by the Siting Commission during the course of screening had strategic planning and a quality assurance program been properly developed and implemented early in the siting process. The

Siting Commission did not develop a strategic plan[2] prior to initiation of screening, and, in fact, to the knowledge of the committee, no such plan was ever developed. The lack of good advance planning made the Siting Commission's task of meeting the already unrealistic deadlines even more difficult to achieve.

Much of the Siting Plan is based on the application of exclusionary factors to yield a number of areas left over from which a site might be chosen. New York's chances of siting an LLRW facility would have improved if from the beginning the Siting Commission had had a clear idea of what was being sought to isolate LLRW. A strategic plan would have forced the Siting Commission to focus on milestones and develop plans to meet them, to develop a project tracking system, and to design a process for developing "work-around" plans to deal with unanticipated problems. The strategic planning process also could have led the commission to consider the need to change its approach at the turning point in the screening process, from one of eliminating unlicensable areas to one of identifying and becoming advocates for licensable sites based on a set of technically grounded siting criteria. The strategic plan also would have forced the Siting Commission to seek out and learn from previous siting efforts. There is a rich literature on this subject that could have been utilized by the commission.

The quality assurance program was not fully developed or implemented when the siting process was halted by the governor in 1990. In fact, the initial draft of the Quality Assurance Plan (Revision 0) was not completed until December 1989. Thus, all of the Siting Commission's screening activities were performed without an approved plan. The procedures that were finally developed came too late to support the important decisions that were made in the early stages of the program.

A quality assurance program would have helped the Siting Commission identify critical data to validate its decisions. The quality assurance plan would have forced the Siting Commission to address several critical aspects of the screening process, including the following:

---

[2]A *strategic plan* is a comprehensive, far-ranging plan for fulfilling an organization's mission over a specified period of time.

- definition of the problem and necessary decisions,
- identification of inputs into the decision process,
- development of decision rules to guide the application of data,
- specifications for data quality, and
- optimal designs for obtaining data.

A quality assurance program also would have helped the Siting Commission implement a document tracking system. Problems with the document tracking system became evident to the committee when it requested a copy of the quality assurance plan and procedures. Although these should have been controlled copies,[3] they could not be located until 10 months after the committee's request.

Not only was the Siting Commission slow in establishing a quality assurance program, it lacked an understanding of the quality assurance program's importance and was unwilling to assume "ownership" responsibility. In response to the committee's question regarding responsibility for implementing the quality assurance program (Appendix F), the commission responded:

> The prime contractor, Weston, was responsible for implementation of the QA Program. The Siting Commission has overall responsibility for establishing a quality assurance program for its own activities and for overseeing its contractors' quality assurance activities.

Implementation of the quality assurance program should have been validated by external audits and surveillance. Only two audits were conducted (for Roy F. Weston, Inc. and its subcontractor, Golder Associates, Inc.). Both were conducted by the staff of Roy F. Weston, Inc., not the Siting Commission.

---

[3]Controlled documents are prepared, reviewed, and approved in accordance with established procedures and are distributed on a controlled basis to ensure that the most current approved information is available to users.

Public Participation

As with most state and federal laws affecting the environment that have been passed since the late 1960s, the 1986 State Act contained specific requirements for public participation. The public was to be informed and their views, comments, information, and analysis heard throughout the siting process. The state law is silent on the basic purposes for such involvement, as are Siting Commission documents.

The committee believes that public involvement in the siting process is important for several reasons. Such involvement is consistent with a democratic form of government founded on the consent of the governed. It can be undertaken to improve decisions by broadening the information base and the range of options and factors considered by decision makers. It can also be undertaken to help ensure that decisions are consistent with public values and preferences and, theoretically at least, are more likely to be supported by the public and, therefore, to be implemented. Experience with siting emphasizes, however, that it is unrealistic to expect that public involvement will reduce conflict where there are fundamental disagreements about goals or that it will necessarily result in solutions to a problem as it is defined by a responsible agency or law.

The early implementation of the public participation plan may have appeared reasonably successful to the Siting Commission in that the public was informed, given an opportunity to provide input before decisions were made, and encouraged to comment on draft plans. As might be expected on the basis of other siting efforts, however, relatively few members of the public availed themselves of the opportunity to provide input at this stage of the process, probably because they did not connect this activity to their own communities. At the same time, there is little evidence that the Siting Commission recognized the difficulties that would be created for later stages of public participation by the set of decisions being put into place during these early stages (e.g., by development of the Siting Plan).

As described in Chapters 3 to 6, problems with public participation began to be much more evident at a particular point in Siting Commission activities: the transition from excluding lands to selecting potential sites. The committee finds that the Siting Commission did not provide the leadership necessary to handle this transition well, probably because it lacked a strategic approach to site selection. The Siting Commission failed to recognize the change from an exclusionary to an inclusive role. That is, the commission did not redefine its role to one of finding a licensable site from one of excluding unlicensable areas. At this transition, public concern and the need for public participation increased significantly, yet critical decisions—for example, the elimination of candidate areas based on a staff decision—appear to have been made by the Siting Commission without adequate formal public review, and many of these decisions were not well documented.

The site selection methodology required a number of subjective decisions by the Siting Commission. These included selecting preference criteria, quantification of scaling criteria, deciding the order in which particular criteria would be applied, choosing an algorithm for scoring, and choosing cutoff scores. The methodology described in the Siting Plan defined a process that calls for careful justification at every step and an ongoing effort to build broad consensus on the many subjective elements of the system. The success (or failure) of such a process would be extremely sensitive to the degree of public participation in Siting Commission activities.

As the discussions in Chapters 5 and 6 illustrate, the level of public participation in these latter stages of screening was insufficient to provide citizens with the necessary level of confidence in the completeness and quality of the Siting Commission's work. Further, much of the public participation in these latter stages was oppositional in nature and focused on previous screening decisions or the legitimacy of the screening process. Yet when confronted with strong public opposition, the Siting Commission chose to continue on with its procedures rather than pause in order to assess the origin of the opposition and provide appropriate responses to the public's questions.

LESSONS TO BE LEARNED

Previous sections of this chapter have focused on the screening process used by the Siting Commission in its attempt to identify sites for an LLRW disposal facility. There are many lessons that could be learned from this effort, in terms of both improving the screening process and siting LLRW disposal facilities generally. In this final section of the report, the committee focuses its attention on the latter issue, that is, the lessons from the New York siting effort that can be applied to LLRW site selection generally.

The screening process employed by the Siting Commission, like the approaches used by many other compacts at the time, is perhaps best characterized as a "top-down" approach in that it entailed a stepwise screening of the entire state to identify ever-smaller parcels of land from which a preferred site could be selected.[4] In addition, the process was designed and executed by a state-level body and its out-of-state consultant with relatively little participation by citizens, particularly during its early stages. Indeed, it was only after areas were designated that significant public interest developed.

There appeared to be relatively few technical or public acceptance problems with the Siting Commission's top-down approach during the early exclusionary phase of the process, probably for two reasons. First, land was being excluded from further consideration, which no doubt pleased citizens of the affected communities. Second, the exclusion decisions were made using good-quality, readily available data that required little in the way of "judgment calls" or interpretation (e.g., boundaries of state and federally protected lands) and that were not in serious dispute by most parties in the process. The limitations of the data became evident, however, once the process passed the "turning point" from exclusion of land to the identification of suitable areas and then of sites—that is, when the siting process shifted from Statewide Exclusionary Screening (SES) (Chapter 4) to CAI (Chapter 5) and PSI (Chapter 6).

The committee believes that there are at least three important lessons to be learned from the Siting Commission's experience. First,

[4]Such an approach can work reasonably well when decisions involve relatively little controversy or where a decision-making body enjoys high levels of public credibility and deference; like may other technically oriented bodies at the time, the Siting Commission may not have been aware that such top-down approaches tend not to work when used for something as controversial as siting a nuclear waste facility.

top-down screening should not be pushed beyond the capabilities of the data and selection criteria to support reasonable and technically defensible decisions. In the case of the New York siting effort, the top-down approach probably should not have been pushed beyond SES, where data were readily available and of reasonably good quality and where the exclusionary criteria were based on laws and regulations that were viewed as reasonable by most parties in the process. The latter stages of screening focused on more abstract, qualitative exercises that required more subjective judgments and that were not as defensible technically. Hence, in the final stages of site identification, the decisions of the Siting Commission seemed arbitrary to many. Attempting to screen out areas based on partial data sets leads to perceptions of unfair treatment, as areas with incomplete descriptive information (often rural regions) are retained for further review, and more thoroughly-described (often urban) areas are excluded, generally to the relief of those living there. Further, the need to create complex data sets to resolve such discrepancies can force the entire siting process toward expensive, time-consuming, and generally unattainable goals.

Once the top-down screening process passes beyond the capabilities of the data and selection criteria to support technically defensible decisions, other strategies need to be considered. A wide variety of options are available, ranging from the collection of additional data to support continued top-down screening to a "volunteer" process. There is an accumulating literature on siting processes that can be used to inform and guide future efforts (e.g., Freudenburg and Rosa, 1984; Downey, 1985; Freudenburg 1985; U.S. Office of Technology Assessment, 1985; Jacob, 1990; Slovic, 1991; Pijawka and Mushkatel, 1992; Dunlap et al. 1993; U.S. Secretary of Energy Advisory Board, 1993; Erikson et al. 1994; Flynn and Slovic 1995; NRC, 1996).[5]

The second lesson to be learned is that public acceptance of the process and its results is necessary to the success of any siting effort. While the committee's charge did not require explicit attention to the public acceptance problems faced by the Siting Commission, it quickly became apparent that—as noted in much of the scientific literature that has now been written on the issue of nuclear waste facility siting (see the

---

[5]Much of this literature has been developed since the Siting Commission began its work in New York State.

previous paragraph for specific references)—the effort to identify waste disposal sites can quickly be derailed because of the failure to devote careful analysis to questions of "sociopolitical" acceptability, which often prove in practice to be impossible to disentangle from questions of "technical" acceptability.

As noted previously, a "superior" site may not be technically superior in all respects but may be a technically acceptable and licensable site that is located in a community that will accept the facility. Public acceptance requires both a technically defensible screening process and a willingness to engage the public substantively on a number of fronts. It is important to recognize, however, that substantive engagement takes time: time to solicit and obtain public input, time to address public perceptions (see the next section), time to make changes in the siting process in response to public comments, and, most of all, time to engage the affected communities effectively at each step of the siting process. As noted previously, time constraints imposed on the New York State Siting Commission by legislative actions and regulations did not allow effective public engagement to occur. Moreover, while substantive public engagement is a necessary ingredient in any siting effort, it is by no means sufficient to ensure success, absent a community willing to accept such a facility.

The third lesson to be learned is that strategic planning is essential to completing complex missions under tight deadlines. Such planning must include the identification of key milestones and objectives as well as procedures for dealing with the many unanticipated problems that inevitably arise in any project of this complexity. To be effective, the strategic plan must be developed at the earliest stages of the process and updated as necessary throughout the project.

Public Perceptions of Nuclear Waste

Nuclear waste, like other toxic wastes, poses significant health risks if not managed properly. Unlike some other waste types, however, the radiation hazard associated with nuclear waste is not literally empirical—that is, the radiation cannot be detected by the (unaided) five senses. Thus, nuclear waste can create what social scientists call an *ambiguity of harm*, and a growing body of empirical research on technological hazards suggests that such an ambiguity of harm may actually create higher levels of measurable stress than do well-understood hazards (for reviews, see Vyner, 1988; Freudenburg and Jones, 1991; Erikson et al., 1994). The nonempirical nature of radioactivity also means that rather than being able to establish a clear sense of controllability over the hazard, the public finds itself dependent on sophisticated radiation monitoring equipment and the technical experts who run that equipment and manage the hazard. While the nuclear industry has managed a complex technology with a high degree of safety for both workers and the public, several highly publicized incidents have occurred where the quality of management and responsibility have been lower than the norm. These incidents have effectively sent to many people a "signal" (Slovic, 1987) that the competence and reliability of experts should not be taken for granted.

As any number of authors have noted, government decisions—specifically including decisions relating to nuclear technology—tended to enjoy relatively widespread public deference in earlier decades, only to become the focus of growing public distrust in recent years. In general terms, the decline is seen as having begun in the mid-1960s, accelerating sharply in the 1970s and continuing into the 1980s (e.g., Lipset and Schneider, 1983); with respect to nuclear technology in particular, the growth in distrust may have begun somewhat later, in the mid-1970s (see, for example, Freudenburg and Baxter, 1985).

In 1984 a committee of the National Research Council (NRC) addressed the effects of public perceptions on nuclear waste facilities. A particularly important emphasis of the committee's report (NRC, 1984) involved what the scientific literature of the day called simply "special effects"—that is, those effects resulting from the presence, or even the potential presence, of a nuclear waste facility in an area. These effects included high levels of public anxiety, opposition, and the potential for stigmatizing regions and peoples. Indeed, the report predicted that "the special effects associated with the radiological mission of the repository will interact with, and may well exceed, the more conventional effects resulting from the location of any large industrial facilities . . ." (NRC, 1984, p. 12).

More recent NRC panels (e.g., NRC, 1993, 1994) have reflected a growing scientific awareness that it is less accurate to refer to "special" impacts than to use the more technical terminology of "opportunity-threat impacts" (see also Freudenburg and Gramling, 1992; Interorganizational Committee on Guidelines and Principles for Socioeconomic Impact Assessment, 1994). As the panel noted, while "changes to physical or biological systems do not occur until a project leads to physical alterations, observable and measurable alterations in the human environment can take place as soon as there are changes in social and economic conditions, which often occur from the time of the earliest rumors or announcements about a project" (NRC, 1994, p. 130). These opportunity-threat impacts grow out of

> the patterns of responses that follow the opportunities and threats that attend proposed development. Speculators buy property, economic development opportunities are created, politicians maneuver for position, interest groups form or redirect their energies, stresses can mount, and a variety of other social and economic effects can occur. . . . These changes have sometimes been called "pre-development" or "anticipatory" impacts, but they are real and measurable. Even the earliest acts of speculators, for example, can drive up the real cost of real estate. (NRC, 1994, p. 131)

The experience of the New York siting effort provides further reinforcement for these conclusions. In the New York effort, there were virtually no on-site impacts. Not a single shovelfull of soil was turned; not a single Siting Commission vehicle or study team was allowed onto the premises of any of the sites. Instead, virtually all of the impacts taking place were the special effects associated with the high levels of public anxiety that, with variations, have also afflicted other efforts to site nuclear waste facilities in this country.

The New York siting effort further confirms the results of social science research and previous experience on siting nuclear waste facilities. These results bear repeating: nuclear waste disposal evokes feelings of anxiety for many members of the public and results in special socioeconomic impacts in the affected communities, and siting efforts ignore these effects at their own peril. Indeed, siting efforts must be structured to address these effects through a high level of public involvement in the siting process.

8

Responses to Questions in Statement of Task

The statement of task for this study (Appendix B) directed the committee to address eight technical questions concerning the screening process used by the Siting Commission to identify potential sites for a low-level radioactive waste (LLRW) disposal facility. These questions deal with a wide range of issues, including the design and application of technical screening criteria and the adequacy of information used by the Siting Commission to make screening decisions.

The committee did not address these questions explicitly in previous chapters of this report mainly because the questions do not match neatly with the various steps of the screening process. Indeed, the questions are quite variable in technical scope, ranging from queries about individual screening criteria to questions about the entire screening process. The committee found it difficult to answer these questions without reviewing the entire screening process.

The committee's review of the screening process in Chapters 2 through 6 of this report is presented roughly in the chronological order in which the siting process was carried out. The purpose of the present chapter is to pull together information from this review to provide explicit responses to the eight questions in the statement of task. The committee provides brief answers to each question below, and it points out where more detailed discussions can be found elsewhere in the report.

QUESTION 1

Are New York State (NYS) exclusionary criteria based on sound technical principles?

Most of the 17 exclusionary criteria developed by the Siting Commission are based on sound principles and address important technical considerations for siting an LLRW disposal facility. As noted in Chapters 5 and 6, however, the committee identified problems with the design or application of three exclusionary criteria. The committee also identified three exclusionary conditions that were not addressed by exclusionary criteria.

- *Criterion 26—air quality nonattainment*—appears to have no basis in legislation or existing regulation, and there is no reason to include it as an exclusionary condition. In the committee's judgment, it should have been designated as a preference criterion.
- *Criterion 4—existing mine exclusion* and *Criterion 44—mineral soil groups*—were applied inappropriately.
- Two preference criteria—*Criterion 20 (erosion)* and *Criterion 3 (subsurface dissolution)*—should have been applied as exclusionary criteria. Additionally, there should have been an exclusionary component to *Criterion 2—seismic hazards.*

QUESTION 2

Was the commission's application of exclusionary criteria to volunteered and non-volunteered lands consistent with good scientific practice?

Exclusionary criteria were applied in the following steps of the screening process: Statewide Exclusionary Screening (SES; see Chapter 4), the Exclusionary Screening and Comparative Preference Analysis steps of Candidate Area Identification (CAI; see Chapter 5), and all steps of Potential Sites Identification (PSI) screening (see Chapter 6). The committee reviewed the application of exclusionary criteria in each of these steps and concluded that screening of nonvolunteered and volunteered lands was generally based on good scientific practice with the following three possible exceptions: (1) application of the exclusionary criteria noted above in the response to Question 1; (2) application of the exclusionary criteria in the Field Surveys step of PSI

(Chapter 6), where not enough information was provided to the committee on how data collected during the surveys were used in site selection; and (3) application of the exclusionary criteria—particularly Criterion 44 (mineral soil groups)—to the offered sites (Chapter 6).

QUESTION 3

Were the commission decisions to exclude sites containing existing mines, which were not part of the exclusionary criteria, based on sound technical considerations?

The committee believes that it was technically reasonable for the Siting Commission to exclude existing mines even though it was not required by the exclusionary criteria or Department of Environmental Conservation (DEC) regulations. As noted in Chapter 5, however, the committee believes that the Siting Commission's rationale for excluding existing mines based on geologic complexity was inappropriate because geologic complexity could not be evaluated properly using the data available to the Siting Commission at the time this criterion was applied. The committee believes that existing mines should have been excluded for a variety of reasons related to their design and suitability for an LLRW disposal facility rather than for geologic complexity.

QUESTION 4

Was the process to select LLRW disposal sites, established by the commission, consistent with good scientific practice?

The process established by the Siting Commission to select LLRW disposal sites, which is outlined in the Siting Plan, was problematical for several reasons. The committee believes that the Siting Commission's use of cutoff scores for preference criteria screening is not justified, because sensitivity analyses (Chapters 5 and 6) suggest a weak correlation between the numerical score and likely site performance. Indeed, the committee's sensitivity analyses suggest that site selection depends to a great extent on the sequence of screening steps and the combination of criteria applied at each step, rather than the likely performance of a site as an LLRW disposal facility.

Additionally, some of the criteria used in screening were not useful for discriminating the desired performance characteristics of sites or were combined with other criteria in inappropriate ways. For example, performance and socioeconomic criteria were combined during CAI and PSI screening. Consequently, a site could receive less than favorable scores on performance criteria and still be selected if it scored highly on socioeconomic criteria. Moreover, many of the performance-related preference criteria were scored using "uncertain" data, that is, data of unknown or uneven accuracy and completeness across the state.

The committee concluded that there was insufficient analysis by the Siting Commission to determine the sensitivity of the scoring system to coupling effects among screening criteria or to small changes in scaling and weighting factors. The committee presents examples of such analyses in Chapters 5 and 6.

QUESTION 5

Were the decisions made by the commission in selecting or narrowing the range of sites for LLRW disposal based on sound technical considerations?

Question 4 addresses the soundness of the screening process used by the Siting Commission to select a site for an LLRW disposal facility. The committee interprets the present question in terms of the manner in which the screening process was applied by the Siting Commission, and particularly the technical soundness of the decisions made by the commission during the various steps of the screening process.

In the committee's judgment, Siting Commission decisions during SES were sound. The exclusionary criteria applied during this step of the screening process were applied appropriately, and the criteria used were based on sound regulatory considerations.

The committee identified several technical problems with Siting Commission decisions during the CAI stage of screening, as described in Chapter 5 of this report. These include the following:

- The use of a cutoff score for Initial Preference Screening is not justified because there is not a good correlation between a numerical score and the likely performance of a site as an LLRW disposal facility.
- The results of screening are nonunique in that the candidate areas selected depend to a great extent on the sequence of screening steps and the combination of criteria applied at each step, rather than the likely performance of an area.
- The sensitivity analysis performed by the Siting Commission was incomplete, poorly documented, and applied to only a small area of the state.

The committee also identified several technical problems with Siting Commission decisions during the PSI stage of the screening process, as noted in Chapter 6 of this report. These included the following:

- The use of a cutoff score in the Geographic Information System (GIS) Screening was not justified because there was not a good correlation between the score and the likely performance of a site.
- The Qualitative Map Assessment step was not discussed in the Siting Plan but was added later to the process. The primary problem with this step is that preference criteria were applied as exclusionary conditions.
- The selection of the five final potential sites was based on a poorly documented staff decision.
- The Siting Commission's decision to include the Taylor North site as one of the five final sites was inconsistent with the commission's requirement that offered sites must be at least as good as other sites. The Taylor North site did not meet the minimum cutoff score applied to other sites at the same stage of screening, and the site contained soils in mineral soil groups 1 through 4, which should have disqualified it from further consideration.

QUESTION 6

Was the development and application of preference criteria (including weighting and scaling factors) designed to identify sites with geological characteristics adequate to contain radioactive waste? Was the commission's use of criteria unrelated to site performance (including weighting and scaling factors) based on good scientific practice?

Four of the 53 preference criteria explicitly address geological characteristics of a disposal site (Appendix I). These are *Criterion 1—geologic complexity*, *Criterion 2—seismic hazards*, *Criterion 3—subsurface dissolution* and *Criterion 5—geologic units*. The committee identified problems with the design or application of two of these criteria, as noted in its response to Question 1. The committee also found the application of Criterion 1 during CAI screening to be inappropriate without more site-specific data. Twelve preference criteria related to natural resources and hydrology (Appendix I) indirectly address the geological characteristics of a disposal site. The committee identified the following problems in the design or application of these criteria:

- The application of *Criteria 12—primary and principal aquifers*—during CAI screening should have accommodated contributed data on aquifer characteristics, well performance, and water usage at the scale of 1-mile-square cells used in CAI screening.
- Data to apply *Criteria 12, 13, and 18—buffers from water resources*—were not available during the stage of the screening process (CAI screening) at which these criteria were first applied. Consequently, a single distance standard (1 mile) was applied across the entire state.
- The design of drainage criteria—particularly *Criterion 14—unconsolidated stratigraphic units*—was unnecessarily restrictive and may have eliminated many otherwise suitable areas of the state from consideration.
- *Criterion 20—erosion*—should have been applied as an exclusionary condition.

• Data to apply *Criterion 22—best usage of surface waters—* were not available at the stage of the screening process (CAI screening) at which this criterion was first applied. Consequently, a single set of distance standards was applied across the entire state.

More detailed discussions of these problems are provided in Chapters 5 and 6 of this report.

As noted in the response to Question 4 above, two aspects of the screening process were not well designed to identify sites with geological characteristics adequate to contain radioactive waste: (1) performance and socioeconomic criteria were combined inappropriately for scoring purposes during CAI and PSI screening; consequently, a site could receive less favorable scores on performance criteria and still be selected if it scored highly on socioeconomic criteria; and (2) scoring of preference criteria related to performance of the site relied on "uncertain" data. Because these criteria address performance factors for an LLRW disposal facility, site scores may be correlated only weakly with the likely performance of the sites.

Most of the nonperformance criteria addressed the impacts on surrounding populations from a disposal facility. Individually, many of these criteria were reasonable; however, the committee believes that they were not well developed or implemented when viewed as a group. As noted in Chapter 6, many of these criteria describe similar or closely coupled phenomena, and their relative weights shifted dramatically through the various stages of site selection. The sensitivity analyses performed by the Siting Commission were inadequate for elucidating the effects of coupling and scoring on selection decisions.

QUESTION 7

Was the commission's application of information obtained through field surveys, site visits, aerial inspections, or any other field investigation used for evaluating potential sites consistent with good scientific practice?

As noted in Chapter 6, the committee reviewed the plan of the field surveys, the training of staff performing the surveys, and the factors to be evaluated. All appear to be conceptually sound.

The windshield surveys conducted during PSI screening were performed over a two-week period. The *Report on Potential Sites Identification* provides windshield survey results only for those sites that were set aside, not for those that were kept. Thus, it was difficult for the committee to assess how the Siting Commission used this information to make its siting decisions.

The data collection forms used in the windshield surveys (see Appendix K) indicate that a broad range of information was to be collected. It is unclear to the committee, however, how this information was actually used in site selection. The committee asked the Siting Commission for, but was never provided, documentation regarding the criteria for the windshield surveys.

The use of the windshield surveys during PSI screening was controversial because of county allegations that information was collected that could have been used to exclude two of the final potential sites (Taylor North and Taylor Central). The survey forms for these sites show that newly constructed incompatible structures were observed close to (and within) the site boundaries. The committee's conclusion in Chapter 6 is that the Siting Commission did not adhere to its stated goals (or to good scientific practice) to evaluate the sites "in light of additional information gained through visual inspection."

QUESTION 8

Did the commission seek and/or utilize information from all available sources, including within identified areas?

As described in Chapter 4, the initial stages of screening were based on readily available information (e.g., from published maps). As the siting process focused on smaller areas of land, the Siting Commission made a reasonable effort to obtain additional information, including information from local governments. Further, the committee believes that the Siting Commission made a sincere effort to review and use this information. Information that was neither used directly nor used to verify information obtained from other sources was deferred for use during independent evaluation of sites rather than for comparison of sites with each other.

The committee believes, however, that the Siting Commission's decision to defer consideration of these data, when viewed in hindsight, was not successful from a public relations point of view, primarily because it appeared to local governments that their offered information was being ignored. The Siting Commission did not communicate effectively to local governments and the public why it decided to defer using much of the information until later stages.

References

Algermissen, S.T., D.M. Perkins, P.C. Thenhaus, S.L. Hanson, and B.L. Bender. 1982. Probabilistic Estimates of Maximum Acceleration and Velocity in Rock in the Contiguous United States. U.S. Geological Survey Open File Report 82-1033.

Coleman, J. 1957. Community Conflict. Glencoe, Ill.: Free Press.

Downey, G. 1985. Politics and technology in repository siting: military versus commercial wastes at WIPP, 1972-1985. Technology in Society 7:47-75.

Dunlap, R.E., M.E. Kraft, and E.A. Rosa. 1993. Public Reactions to Nuclear Waste: Citizens' Views of Repository Siting. Durham, N.C.: Duke University Press.

Erikson, K.T., E.W. Colglazier, and G.F. White. 1994. Nuclear waste's human dimension. Forum for Applied Research and Public Policy 9(3): 91-97.

Flynn, J., and P. Slovic. 1995. Yucca Mountain, a crisis for policy: prospects for America's high-level nuclear waste program. Annual Review of Energy and the Environment 20:83-118.

Freudenburg, W.R. 1985. Waste not: the special impacts of nuclear waste facilities. Pp. 75-80 in J.G. McCray et al., eds., Waste Isolation in the U.S., Vol. 3: Waste Policies and Programs. Tucson: Univ. of Arizona Press.

Freudenburg, W.R., and R.K. Baxter. 1984. Host community attitudes toward nuclear power plants: A reassessment. Social Science Quarterly 65(4):1129-1136.

Freudenburg, W.R., and R. Gramling. 1994. Oil in Troubled Waters: Perceptions, Politics, and the Battle over Offshore Oil. Albany: State University of New York (SUNY) Press.

Freudenburg, W.R., and T.R. Jones. 1991. Attitudes and stress in the presence of technological risk: a test of the Supreme Court hypothesis. Social Forces 69(4):1143-1168.

Freudenburg, W.R., and S.K. Pastor. 1992. Public responses to technological risks: toward a sociological perspective. Sociological Quarterly 33(3):389-412.

Freudenburg, W.R., and E.A. Rosa, eds. 1984. Public Reactions to Nuclear Power: Are There Critical Masses? Boulder, Colo.: American Association for the Advancement of Science/Westview.

Galanter, M. 1974. Why the 'haves' come out ahead: Speculations on the limits of legal change. Law and Society Review 9:95-160.

Interorganizational Committee on Guidelines and Principles for Socioeconomic Impact Assessment. 1994. Guidelines and Principles for Socioeconomic Impact Assessment. Washington, D.C.: U.S. National Marine Fisheries Service.

Jacob, G. 1990. Site Unseen: The Politics of Siting a Nuclear Waste Repository. Pittsburgh: Univ. of Pittsburgh Press.

Kunreuther, H., K. Fitzgerald, and T.D. Aarts. 1993. Siting noxious facilities: a test of the facility siting credo. Risk Analysis 13(3):301-318.

League of Women Voters. 1993. The Nuclear Waste Primer: A Handbook for Citizens. New York: Lyons & Burford.

Licensing Requirements for Land Disposal of Radioactive Waste. Code of Federal Regulations, Title 10, Part 61 (10 CFR 61). 1995 edition.

Lipset, S.M., and W. Schneider. 1983. The decline of confidence in American institutions. Political Science Quarterly 98:379-402.

Low-Level Radioactive Waste Forum. 1996. Summary Report: Low-Level Radioactive Waste Management Activities in States and Compacts. A Supplement to LLW Notes 4(1). Washington, D.C.

Low-Level Radioactive Waste Policy Act. 1980 (P.L. 96-573, December 23). United States Statutes at Large 94:3347-3349.

Low-Level Radioactive Waste Policy Amendments Act of 1985. 1986 (P.L. 99-240, Title I, January 15, 1986). United States Statutes at Large 99:1842-1859.

McCray, J.G. et al., eds. Waste Isolation in the U.S. Vol. 3: Waste Policies and Programs. Tucson: Univ. of Arizona Press.

NRC (National Research Council). 1984. Social and Economic Aspects of Radioactive Waste Disposal: Considerations for Institutional Management. Washington, D.C.: National Academy Press.

NRC. 1990. Ground Water Models: Scientific and Regulatory Applications. Washington, D.C.: National Academy Press.

NRC. 1992. Assessment of the U.S. Outer Continental Shelf Environmental Studies Program: III. Social and Economic Studies. Washington, D.C.: National Academy Press.

NRC. 1993. Assessment of the U.S. Outer Continental Shelf Environmental Studies Program: IV. Lessons and Opportunities. Washington, D.C.: National Academy Press.

NRC. 1994. Environmental Information for Outer Continental Shelf Oil and Gas Decisions in Alaska. Washington, D.C.: National Academy Press.

NRC. 1996. Understanding Risk: Informing Decisions in a Democratic Society. Washington, D.C.: National Academy Press.

New York State Department of Environmental Conservation, Bureau of Energy and Radiation. 1987. Final Environmental Impact Statement for Promulgation of 6 NYCRR Part 382: Regulations for Low-Level Radioactive Waste Disposal Facilities (Certification of Proposed Sites and Disposal Methods). 2 vols. Albany, N.Y.

New York State Department of Environmental Conservation, Bureau of Radiation. 1993. 6 NYCRR 382: Regulation of Low-Level Radioactive Waste (LLRW) Disposal Facilities: Certification of Proposed Site and Disposal Methods. New York Code of Rules and Regulations, Title 6, Part 382. (March 14.) Albany, N.Y.

New York State Legislature. 1986. Low-Level Radioactive Waste Management Act. Laws of the State of New York, Chapter 673 (July 26):2801-2817.

New York State Legislature. 1990. An Act to Amend the Environmental Conservation Law in Relation to the Membership and Duties of the Commission . . . and the Advisory Committee . . . [amendment to 1986 LLRW Management Act]. 1990 Session Laws, Chapter 913 (A. 12080, July 30).

New York State Low-Level Radioactive Waste Siting Commission. 1988a. Candidate Area Identification Report. Albany, N.Y.

New York State Low-Level Radioactive Waste Siting Commission. 1988b. Plan for Selecting Sites for Disposal of Low-Level Radioactive Wastes. Albany, N.Y.

New York State Low-Level Radioactive Waste Siting Commission. 1988c. Statewide Exclusionary Screening Report. Albany, N.Y.

New York State Low-Level Radioactive Waste Siting Commission. 1989a. Disposal Method Screening Report. Albany, N.Y.

New York State Low-Level Radioactive Waste Siting Commission. 1989b. Report on Potential Sites Identification (includes ten volumes of appendixes). Albany, N.Y.

New York State Low-Level Radioactive Waste Siting Commission. 1993. Excluded Areas Report. Albany, N.Y.

New York State Low-Level Radioactive Waste Siting Commission. 1995. 1994 Source Term Report Executive Summary. Albany, N.Y.

Pijawka, D.K., and A.H. Mushkatel. 1992. Public opposition to the siting of the high-level nuclear waste repository: the importance of trust. Policy Studies Review 10(4): 180-194.

Slovic, P. 1987. Perception of risk. Science 236:280-285.

Slovic, P. 1991. Perceived risk, trust, and the politics of nuclear waste. Science 254:1603-1607.

Sullivan, T.M., and M. Chehata. 1995. Overview of Research and Development in Subsurface Fate and Transport Modeling (BNL-52469). Upton, New York: Brookhaven National Laboratory.

U.S. General Accounting Office. 1992. Nuclear Waste: New York's Adherence to Site Selection Procedures Is Unclear. Report prepared for the Honorable Alfonse M. D'Amato, U.S. Senate (GAO/RCED-92-172). Washington, D.C.

USNRC (U.S. Nuclear Regulatory Commission). 1989. Quality Assurance Guidance for Low-Level Radioactive Waste Disposal Facility (NUREG-1293). Washington, D.C.: U.S. Nuclear Regulatory Commission.

U.S. Office of Technology Assessment. 1985. Managing the Nation's Commercial High-Level Radioactive Waste. Washington, D.C.: U.S. Government Printing Office.

U.S. Secretary of Energy Advisory Board, Task Force on Radioactive Waste Management. 1993. Earning Public Trust and Confidence: Requisites for Managing Radioactive Wastes. Washington, D.C.: U.S. Department of Energy.

Vyner, H.M. 1988. Invisible Trauma: The Psychosocial Effects of Invisible Environmental Contaminants. Lexington, Mass.: Lexington Books.

Appendixes

Appendix A

List of Acronyms and Abbreviations

1980 Act	Federal Low-Level Radioactive Waste Policy Act of 1980 (P.L. 95-573)
1985 Amendments Act	Federal Low-Level Radioactive Waste Policy Amendments Act of 1985 (P.L. 99-240)
1986 State Act	New York State Low-Level Radioactive Waste Management Act
Advisory Committee	New York State Low-Level Radioactive Waste Advisory Committee
BRWM	Board on Radioactive Waste Management
CAI	Candidate Area Identification
CAIR	New York State Low-Level Radioactive Waste Siting Commission's *Candidate Area Identification Report*
DEC	New York State Department of Environmental Conservation
ECL	New York State *Environmental Conservation Law*
EPA	U.S. Environmental Protection Agency
FEMA	Federal Emergency Management Agency
GIS	Geographic Information System
LLRW	Low-level radioactive waste
NORM	Naturally occurring radioactive materials
NRC	National Research Council
NYS	New York State
PSI	Potential Sites Identification
ROPSI	New York State Low-Level Radioactive Waste Siting Commission's *Report on Potential Sites Identification*
SES	Statewide Exclusionary Screening

SESR	New York State Low-Level Radioactive Waste Siting Commission's *Statewide Exclusionary Screening Report*
Siting Commission	New York State Commission for Siting a Low-Level Radioactive Waste Facility
Siting Plan	New York State Low-Level Radioactive Waste Siting Commission's *Plan for Selecting Sites for Disposal of Low-Level Radioactive Wastes*
U.S.C.	*U.S. Code*
USNRC	U.S. Nuclear Regulatory Commission

Appendix B

Statement of Task

BRWM [Board on Radioactive Waste Management] will undertake a task-oriented examination of the scientific, technical, and procedural issues in siting a low-level radioactive waste (LLRW) facility in New York State, and in choosing the disposal or isolation technology for such a site. A committee of the board will evaluate the specific reports and documentation provided by NYS Department of Health (DOH) relating to the scientific, technical, and procedural approach used by the Siting Commission in (1) the site selection process and excluding lands from consideration, (2) selection of a preferred disposal technology, and (3) projection of the LLRW source term. In addition, the committee will consider scientific and technical material from all interested parties.

The NRC/BRWM will arrange for the empanelment of a multidisciplinary scientific and technical review committee of about 18 experts in appropriate fields such as geology, hydrology, geochemistry, seismology, civil and mining engineering, material science, toxicology, LLRW processing and management, risk analysis, public health, environmental law, health physics, and quality assurance. The committee will meet as a committee of the whole and as three subcommittees on siting, disposal methodology, and source term determination.

The first three tasks in a five-task study will include the study of documents related to the criteria, methodology, procedures, and the decision process used by the Siting Commission to select potential LLRW facility sites. The review will evaluate the nature, sources, and quality of any specific data, analyses, procedures, modeling, and calculations that the commission used in the decision-making process. These documents will relate to (1) considerations in the exclusion of lands as potential sites, (2) how the range of alternative sites was narrowed to 10 candidate areas, and (3) how the 5 potential sites were selected.

The fourth task will review commission reports and other information relating to the selection and justification of a preferred disposal method. The same considerations will be evaluated as for the siting documents.

The fifth task will address the projected source term. The committee will review commission reports and consider other scientific and technical information relating to the decision-making process leading to projection of the types, quantities, or activity level of LLRW, or source term, intended for emplacement in a permanent LLRW facility. The same considerations will be evaluated as for the siting and the disposal technology documents.

The results of the study will be five reports, one for each of the tasks, and, if feasible, a final summary report which will be reviewed and distributed in accordance with NAS-NRC procedures. NYS DOH will receive 250 as per their request; additional copies will be provided to BRWM committee members and other parties in accordance with NRC policy. The reports will be made available upon request to all states and the public without restriction.

As part of its review the committee will address the following questions and concerns of importance to New York State:

a. Are NYS exclusionary criteria based on sound technical principles?

b. Was the commission's application of exclusionary criteria to volunteered and nonvolunteered lands consistent with good scientific practice?

c. Were the commission's decisions to exclude sites containing existing mines, which were not part of the exclusionary criteria, based on sound technical considerations?

d. Was the process to select LLRW disposal sites, established by the commission, consistent with good scientific practice?

e. Were the decisions made by the commission in selecting or narrowing the range of sites for LLRW disposal based on sound technical considerations?

f. Was the development and application of preference criteria (including weighting and scaling factors) designed to identify sites with geological characteristics adequate to contain radioactive waste? Was the commission's use of criteria unrelated to site performance (including weighting and scaling factors) based on good scientific practice?

g. Was the commission's application of information obtained through field surveys, site visits, aerial inspections, or any other field investigation used for evaluating potential sites consistent with good scientific practice?

h. Did the commission seek and/or utilize information from all available sources, including within identified areas?

Where questions posed raise policy concerns as well as scientific ones, the committee will attempt to be clear about these distinctions. The committee will take no position on policy matters that it identifies.

Appendix C

Biographical Sketches of Committee Members

Susan D. Wiltshire, *Chair*, is vice president of JK Research Associates, Inc., in Hamilton, Massachusetts, consulting in the areas of radioactive waste management, public involvement in policy and technical decisions, and risk communication. Ms. Wiltshire has served on a number of National Research Council committees, including the Board on Radioactive Waste Management. She is a former chairman of the elected Board of Selectmen, chief executive body of the town of Hamilton, former president of the League of Women Voters of Massachusetts, and current chairman of the board of Northeast Health Systems, Inc. Ms. Wiltshire wrote the 1993 U.S. League of Women Voters publication *A Nuclear Waste Primer* and serves on the League's Nuclear Waste Education Project Planning Committee. She received a B.Sc. degree in mathematics from the University of Florida and has been trained as a mediator.

Robert J. Ahrens is lead scientist, soil taxonomy, for the Natural Resources Conservation Service (formerly the Soil Conservation Service), U.S. Department of Agriculture. He has extensive experience in mapping soils and is an expert in soil genesis, morphology, and soil classification. His research activities center around the genesis of soils derived from the tephras of Mount St. Helens. He earned his Ph.D. in agronomy (soil science) from the University of Nebraska, Lincoln.

Gloria Anderson has been an active member of the League of Women Voters in California for 30 years. She was natural resources director of the League of Women Voters of California for four years. During the past nine years, she has worked on public participation projects relating to low-level radioactive waste disposal facility siting in California and on hazardous waste management planning in San Bernardino County. She has authored public participation guidebooks published by the league and a manual for local assessment committees involved in the siting of specified hazardous waste facilities. She received a B.S. in secondary education from the University of Wisconsin-Eau Claire and an M.S. in speech from the University of Wisconsin-Madison.

Charles A. Baskerville is professor of geology in the Department of Physics/Earth Sciences at Central Connecticut State University. He retired from the U.S. Geological Survey as a project chief specializing in urban engineering, geologic mapping, and research in landslide processes. He was a dean and professor of engineering geology at the City University of New York. He served as commonwealth visiting professor of geology at George Mason University. He has served on several panels of the National Research Council (NRC) and as consultant to the New York State Department of Environmental Protection's Water Tunnel Project. He received a Ph.D. in geology from New York University.

Randy L. Bassett is a professor of environmental geochemistry in the Department of Hydrology and Water Resources at the University of Arizona. His research interests include stable isotope geochemistry, chemical modeling, flow and transport of contaminants, and mass transfer geochemistry. He is currently on the editorial review board for *Groundwater*. He was on the board of directors of the Association of Ground Water Scientists and Engineers and was honored as Darcy distinguished lecturer. Dr. Bassett presently serves on the Scientific and Industrial Advisory Committee for the Waterloo Centre for Groundwater Research. He formerly served on the U.S. Environmental Protection Agency (EPA) Proposal Review Panel and on a Radioactive Waste Panel for Argonne National Laboratory. He received a Ph.D. in geochemistry from Stanford University in 1977.

Lynda L. Brothers is a partner in the Seattle office of the national law firm of Davis Wright Tremaine and specializes in environmental, natural resource, energy, and administrative law. She received a B.S. in genetics from the University of California, Berkeley, and an M.S. in biology from the University of Virginia, Charlottesville. She was deputy assistant secretary, U.S. Department of Energy (DOE), 1978-1980, and assistant director, Washington Department of Ecology, prior to entering private practice. Her law practice deals with the regulation, transportation, and disposal of radioactive, hazardous, and solid wastes as well as the regulation of water and air emissions.

Halina Brown is professor of environmental health and chair of the Environmental Science and Policy Program at Clark University. Her present research interests include the use of science in public policy, the effects of pollution and industrial development in Poland on cancer mortality, U.S. and European environmental health policy, and corporate hazard management in developing countries. Prior to joining Clark University, she was a chief toxicologist for the Massachusetts Department of Environmental Protection. Dr. Brown has served on numerous state and national advisory panels, including most recently the National Research Council Committee to Review Risk Management in the DOE's Environmental Restoration Program, Massachusetts Toxic Use Reduction Science Advisory Board, and National Science Foundation proposal review panels. She is a fellow of the Society for Risk Analysis. She received her Ph.D. in chemistry from New York University in 1975.

Gail A. Cederberg is a consulting environmental engineer specializing in hydrology. Her work focuses on hazardous waste investigations, design and implementation of remedial actions, modeling ground water flow and geochemical transport, and regulatory policy and analysis. She developed TRANQL, a finite element coupled transport-geochemical reaction model, to investigate the transport of metals in ground water systems. She served as the chairperson and peer reviewer for the DOE Office of Health and Environmental Research Review's Panels on Field Research Modeling and Multicomponent Predictive Modeling Subprograms. She was a member of a DOE Subsurface Science Program, Workshop Advisory Panel. Dr. Cederberg received her Ph.D. in civil engineering (hydrology and water resources specialty) from Stanford University.

John L. Croes is an assistant vice president/senior program manager with Science Applications International Corporation. He has chaired several international environmental and waste management conferences that have had major impacts on approaches now taken by the industry. He has been involved with many large programs, including the DOE Office of Civilian Radioactive Waste Management's High-Level Waste Program, and has helped develop quality assurance programs for various other waste management programs. He holds an engineering degree from Tennessee Technological University and spent several years on the Apollo and Skylab programs. He has received many awards, including the 1994 Person of the Year award by the Energy and Environmental Division of the American Society of Quality Control.

William P. Dornsife is director of the Bureau of Radiation Protection at the Pennsylvania Department of Environmental Resources. In this position he is responsible for providing overall planning, direction, and implementation of all state radiation programs. He sits on the Executive Committee of the Low-Level Radioactive Waste (LLRW) Forum and is chairperson-elect of the Conference of Radiation Control Program Directors and a full member of the National Council on Radiation Protection and Measurements. He also chairs the U.S. EPA

National Advisory Committee on Environmental Policy and Technology's Subcommittee on Radiation Cleanup Regulation. Mr. Dornsife is a registered professional engineer in Pennsylvania, and a graduate of the U.S. Naval Academy. He received his M.S. in nuclear engineering from Ohio State University.

John E. Ebel is a professor of geophysics at Boston College and director of Weston Observatory of Boston College. A specialist in earthquake seismology and seismic hazards, he supervises the operation of a regional seismic network in New England to monitor earthquake activity in the northeastern United States. He is a former chairman of the Eastern Section of the Seismological Society of America and is editor-in-chief of *Seismological Research Letters*. He received his Ph.D. in geophysics from the California Institute of Technology.

William R. Freudenburg is a professor of rural sociology and environmental studies at the University of Wisconsin-Madison. He is a specialist on the human aspects of risk assessment and risk management and has done extensive research on nuclear and other energy technologies. He currently chairs Section K (Social, Economic, and Political Sciences) of the American Association for the Advancement of Science. He has served on several NRC committees and federal advisory committees relating to energy and waste management issues. He was the first congressional fellow from the American Sociological Association to serve in the U.S. House of Representatives. Dr. Freudenburg received his Ph.D. in sociology from Yale University in 1979.

Robert D. Hatcher, Jr., is distinguished scientist and professor of geology at the University of Tennessee and Oak Ridge National Laboratory. He was given the first Geological Society of America Distinguished Service Award in 1988 after serving as editor of the *GSA Bulletin.* He has served on and presided over many committees for the federal, state, and international communities. He was president of the Geological Society of America. Currently, he is serving as president of the American Geological Institute, member of the National Research Council's Board on Radioactive Waste Management, member of the Nuclear Regulatory Commission's Nuclear Reactor Safety Research Review Committee, and codirector of the Institute of Geotechnology at the University of Tennessee. His primary research is directed toward the evolution of continental crust through the formation of mountain chains. He received his Ph.D. in structural geology at the University of Tennessee.

Carol Hornibrook manages the Electric Power Research Institute's low-level waste and radiation protection research. The program focuses on minimizing the volumes of waste generated at nuclear power plants, using advanced technologies to optimize liquid waste processing and better define the impacts of low-level waste disposal. In addition, Ms. Hornibrook presents information on utility low-level waste issues at courses offered at both the Massachusetts Institute of Technology and the Harvard School of Public Health. She received her B.S. in biology from the State University of New York, Plattsburgh, and an M.S. degree in urban and environmental planning from Rennsalaer Polytechnic Institute.

Janet A. Johnson is a senior radiation scientist with Shepherd Miller, Inc., with more than 30 years of experience in radiation protection. She has a B.S. degree in chemistry, M.S. in radiation biology, and Ph.D. in environmental health. Dr. Johnson is a certified health physicist and a certified industrial hygienist. Her principal areas of expertise include environmental radiation risk assessment, waste management, and radiation protection training. She currently serves on the Colorado Hazardous Waste Commission and the Governor's Radiation Advisory Board and is a member of the Radiation Advisory Committee of the EPA's Science Advisory Board.

Linda L. Lehman is the president of L. Lehman & Associates, Inc., a hydrogeologic consulting firm located in Minneapolis, Minnesota, and Moscow, Russia. Most of her work involves technical and regulatory analysis of nuclear and hazardous waste disposal facilities. She has been a consultant to several states, compacts, Indian nations, the Commission of European Communities, major oil companies, and the nuclear industry on nuclear or environmental regulation in the United States and Russia. She received an M.S. degree in hydrogeology from the University of South Florida.

Robert Meyer is vice president of Keystone Scientific, Inc. In this capacity he is involved in radionuclide and chemical dose reconstructions for the Rocky Flats nuclear weapons facility, near Denver, Colorado, and the Savannah River Site near Aiken, South Carolina. Dr. Meyer was vice president in charge of Chem-Nuclear Systems' contract with Pennsylvania from 1990 to 1992 and was responsible for initial efforts to site the Appalachian Compact's LLW facility. He has served on the Radiation Advisory Committee of the EPA's Science Advisory Board and is currently a member of the National Research Council's Board on Radioactive Waste Management. He has also served as a consultant to the International Atomic Energy Agency in Vienna, coauthoring a report on cleanup after accidental releases of radioactive materials. Dr. Meyer received his Ph.D. in radiation biology from Colorado State University.

Della M. Roy is professor emerita of materials science at the Materials Research Laboratory and Department of Materials Science and Engineering at Pennsylvania State University. Her research interests include materials synthesis and characterization in inorganic, ceramic, cement, and mineral systems and nuclear and chemical waste management. She is the founding editor and editor-in-chief of *Cement and Concrete Research.* She has been very active in professional associations and is a fellow of the American Ceramic Society, the American Concrete Institute, the Mineralogical Society of America, and the American Association for the Advancement of Science. Dr. Roy is a member of the National Academy of Engineering and has served on several committees, including the president's committee. She received her Ph.D. in mineralogy from Pennsylvania State University.

Miklos D.G. Salamon is professor of mining engineering and past head of the Department of Mining Engineering at the Colorado School of Mines. He has been a mining engineer and has directed research into safety in mines and the development of advanced mining technology for over 40 years. Dr. Salamon has held various research and/or teaching appointments in England, South Africa, Australia, and the U.S. He is an international consultant specializing in mining technology and mine safety, and has participated in the investigation of several major mining disasters. He was a member of the Commission of Inquiry into Safety and Health in the Mining Industry that submitted its report to the president of the Republic of South Africa in 1995. Dr. Salamon received his Ph.D. in mining engineering from King's College, University of Durham, United Kingdom.

Leonard C. Slosky is president of Slosky and Company, an environmental consulting firm in Denver, Colorado. He specializes in environmental site assessments, subsurface investigations, site remediation, human health risk assessment, and hazardous and radioactive waste management. He has served as a consultant to the state of Nevada and Clark County, Nevada, on high-level nuclear waste. He has conducted independent reviews of the Rocky Flats Plant (Plutonium Vulnerability Assessment) and the Waste Isolation Pilot Project (WIPP Blue Ribbon Panel). He has been executive director of the Rocky Mountain Low-Level Radioactive Waste Board since 1983. He received a B.A. in environmental technology assessment from the University of Colorado, Boulder.

Arthur A. Socolow is presently a consulting geologist specializing in environmental subjects, including site evaluation, mineral resource issues, and waste disposal projects. A former professor of geology, his background includes service with the U.S. Geological Survey and three decades as state geologist and director of the Pennsylvania Geological Survey. A fellow of the Geological Society of America and the Society of Economic Geologists, he has also served as president of the Association of American State Geologists, chairman of the American Association for the Advancement of Science's Geology Section, chairman of the Interstate Oil Compact Commission's Research Committee, chairman of the American Commission on Stratigraphic Nomenclature, and member of the Outer Continental Policy Committee of the U.S. Department of Interior. He received his Ph.D. in geology from Columbia University.

Appendix D

List of Presenters to the Committee

Cynthia Banas, Resident, Vernon, New York; League of Women Voters

Fred Bateman, Don't Waste New York—Central counties

Bob Bates, Don't Waste New York—Western counties

Duane Chapman, Professor of Economic Resources, Cornell University

Denise Cote-Hopkins, Coordinator, Cortland County Low-Level Radioactive Waste Office

Fay Cougakis, Citizen

Barry Crandall, Resident, Taylor County

Karen Crandall, Resident, Taylor County

Alfonse D'Amato, U.S. Senator from New York, (written submission)

Jamie Dangler, Assistant Professor of Sociology, State University of New York-Cortland; United University Professions; New York State United Teachers

Mary Ann Diaz, Resident, Taylor County

Louise Dressen, Environmental Issues Management, Consultant to the New York State Siting Commission

Jay Dunkelberger, Technical Director, New York State Siting Commission

David Gage, Earth Day Southern Tier, Binghamton, New York; Energy Fair, Cortland, New York

Cindy Gagne, Resident, Oswego County

David Gerber, President, Protect Our World's Environmental Resources (P.O.W.E.R.)

Eugene Gleason, Deputy Commissioner for Operations, New York State Energy Office

William Griffen, Ecology Action of Cortland County; State University of New York-Cortland

Susan Griffin, Coordinator, Chenango North Energy Awareness

David Harris, Citizens Advisory Committee

Nancy Jarvis, Coordinator of Ground-Water Management, Cortland County Planning

Pamela Jenkins, Resident, Cortland County

Gordon I. Kaye, Chairman, New York State Low-Level Waste Group

Dooley Kieffer, Copy Editor, Ecological Society of America; Resident, Cortland County

Beverly Livesay, Tompkins County Board of Representatives

Alan Lum, Retired citizen

John Matuszek, New York State Department of Health

Paul Merges, Chief, Bureau of Radiation, New York State Department of Environmental Conservation (DEC), on behalf of Commissioner Thomas Jorling, and Edward Sullivan, Deputy Commissioner for Environmental Quality, DEC

Paul Meyers, for Ned Sullivan, DEC

Todd Miller, U.S. Geological Survey

Cindy Monaco, Consultant to the committee

Mark Monmonier, Professor of Geography, Syracuse University

Angelo Orazio, Chairman, New York State Siting Commission

Marian Rose, Sierra Club

Katherine Shelly, Chair, Nuclear Task Force, Return the Environment of Susquehanna Country Under Ecology (RESCUE)

William Stasiuk, Director, Center for Environmental Health, New York State Department of Health

Richard E. Tupper, Chairman, Cortland County Legislature

F. William Valentino, New York State Environmental Research and Development Authority

Lori Wooten, Protect Our World's Environmental Resources (P.O.W.E.R.)

Appendix E

List of Materials Reviewed

Ahlström, P-E. 1994. Sweden takes steps to solve waste problem. Forum for Applied Research and Public Policy 9(3):119-121.

Algermissen, S.T., D.M. Perkins, P.C. Thenhaus, S.L. Hanson, and B.L. Bender. 1982. Probabilistic Estimates of Maximum Acceleration and Velocity in Rock in the Contiguous United States. U.S. Geological Survey Open File Report 82-1033.

American Society of Mechanical Engineers. 1986. Quality Assurance Program Requirements for Nuclear Facilities, Supplement 17S-1 (NQA-1).

Blanchard, M.B., D.C. Dobson, M.D. Voegele, and J.L. Younker. 1991. Licensing a geologic repository: types of uncertainties. Environmental Geology and Water Sciences 18(3):215-220.

Brown, P., and E.J. Mikkelsen. 1990. No Safe Place: Toxic Waste, Leukemia, and Community Action. Berkeley: University of California Press.

Chess, C., B.J. Hance, and P.M. Sandman. n.d. Improving Dialogue with Communities: A Short Guide for Government Risk Communication. Report submitted to New Jersey Department of Environmental Protection, Division of Science and Research.

Coleman, J. 1957. Community Conflict. Glencoe, Ill.: Free Press.

Cortland County Low-Level Radioactive Waste Office. 1990. First Annual Report: New York State's Low-Level Radioactive Waste Program—The Cortland Experience.

Cortland County Low-Level Radioactive Waste Office. 1991. Presentation to the New York State Low-Level Radioactive Waste Siting Commission: Outstanding issues regarding the NYS LLRW Siting Commission's Potential Sites Identification process, presented by C. Monaco, September 11, Albany, N.Y.

Cortland County Low-Level Radioactive Waste Office. 1991. Presentation to the New York State Low-Level Radioactive Waste Siting Commission Regarding Its Site Selection Plan Procedures, presented by C. Monaco, July 19, Albany, N.Y.

Cortland County Low-Level Radioactive Waste Office. 1994. Correspondence to J. Croes from D. Cote-Hopkins, November 29, transmitting copy of correspondence to M. Monmonier from D. Cote-Hopkins, September 19, regarding concerns about Preference Criteria 51, Proximity to Waste Generators.

Cortland County Low-Level Radioactive Waste Office. 1994. Correspondence to L. Lehman from D. Cote-Hopkins, November 29, transmitting correspondence to K. Abbey from J. Randall, New York State Low-Level Waste Siting Commission, May 13, 1992, regarding response to question concerning Taylor North's position in list of potential sites.

Covello, V. 1993. Risky business. Nuclear Energy First Quarter:8-9.

Creighton, J.L. 1980. Public Involvement Manual: Involving the Public in Water Power Resources Decisions. Washington, D.C.: U.S. Government Printing Office.

Dietz, T., P.C. Stern, and R.W. Rycroft. 1989. Definitions of conflict and the legitimation of resources: the case of environmental risk. Sociological Forum 4(1):47-70.

Downey, G. 1985. Politics and technology in repository siting: military versus commercial wastes at WIPP, 1972-1985. Technology in Society 7:47-75.

Dunlap, R.E., M.E. Kraft, and E.A. Rosa. 1993. Public Reactions to Nuclear Waste: Citizens' Views of Repository Siting. Durham, N.C.: Duke University Press.

English, M.R., and E.W. Colglazier. 1994. Role of politics in waste dispute. Forum for Applied Research and Public Policy 9(3):72-79.

Erikson, K., E.W. Colglazier, and G.F. White. 1994. Nuclear waste's human dimension. Forum for Applied Research and Public Policy 9(3):91-97.

Farley, J.M. 1994. Waste-siting tussle: a utility perspective. Forum for Applied Research and Public Policy 9(3):107-111.

Fischhoff, B., S. Lichtenstein, P. Slovic, S.L. Derby, and R.L. Keeney. 1981. Acceptable Risk. New York: Cambridge University Press.

Flynn, J., W. Burns, C.K. Mertz, and P. Slovic. 1992. Trust as a determinant of opposition to a high-level radioactive waste repository: analysis of a structural model. Risk Analysis 12(3):417-430.

Flynn, J., and P. Slovic. 1995. Yucca Mountain, a crisis for policy: prospects for America's high-level nuclear waste program. Annual Review of Energy and the Environment 20:83-118.

Flynn, J., P. Slovic, and C.K. Mertz. 1984. Gender, race, and perception of environmental health risks. Unpublished paper, Eugene, Ore.

Frank, J. 1932. Mr. Justice Holmes and non-Euclidean legal thinking. Cornell Law Quarterly 17:568-603.

Freudenburg, W.R. 1985. Waste not: the special impacts of nuclear waste facilities. Pp. 75-80 in J.G. McCray et al., eds., Waste Isolation in the U.S., Vol. 3: Waste Policies and Programs. Tucson: University of Arizona Press.

Freudenburg, W.R. 1988. Perceived risk, real risk: social science and the art of probabilistic risk assessment. Science 242(Oct. 7):44-49.

Freudenburg, W.R. 1992. Heuristics, biases, and the not-so-general publics: expertise and error in the assessment of risks. Pp. 229-249 in D. Golding and S. Krimsky, eds., Theories of Risk. Westport, Conn.: Greenwood.

Freudenburg, W.R. 1993. Risk and recreancy: Weber, the division of labor, and the rationality of risk perceptions. Social Forces 71(4):909-932.

Freudenburg, W.R., and R.K. Baxter. 1984. Host community attitudes toward nuclear power plants: a reassessment. Social Science Quarterly 65(4):1129-1136.

Freudenburg, W.R., and R. Gramling. 1994. Oil in Troubled Waters: Perceptions, Politics, and the Battle over Offshore Oil. Albany: State University of New York (SUNY) Press.

Freudenburg, W.R., and T.R. Jones. 1991. Attitudes and stress in the presence of technological risk: a test of the Supreme Court hypothesis. Social Forces 69(4):1143-1168.

Freudenburg, W.R., and S.K. Pastor. 1992. Public responses to technological risks: toward a sociological perspective. Sociological Quarterly 33(3):389-412.

Freudenburg, W.R., and E.A. Rosa, eds. 1984. Public Reactions to Nuclear Power: Are There Critical Masses? Boulder, Colo.: American Association for the Advancement of Science/Westview.

Galanter, M. 1974. Why the 'haves' come out ahead: Speculations on the limits of legal change. Law and Society Review 9:95-160.

Gertz, C.P., and S. Teitelbaum. 1994. Yucca Mountain reveals its secrets to scientists. Forum for Applied Research and Public Policy 9(3):106-112.

Golding, D., and S. Krimsky, eds. 1992. Theories of Risk. Westport, Conn.: Greenwood.

Harris, W.E. 1993. Siting a hazardous waste facility: a success story in retrospect. Risk Analysis 13(3):3-4.

Hatcher, R. 1994. Correspondence to I. Alterman, April 4, transmitting handouts from March 28, presentation by New York State Low-Level Radioactive Waste Siting Commission at Northeast Geological Society of America meeting: "How do you find the 'best' site for a disposal facility nobody seems to want?"

Interorganizational Committee on Guidelines and Principles for Socioeconomic Impact Assessment. 1994. Guidelines and Principles for Socioeconomic Impact Assessment. Washington, D.C.: U.S. National Marine Fisheries Service.

Jacob, G. 1990. Site Unseen: The Politics of Siting a Nuclear Waste Repository. Pittsburgh: University of Pittsburgh Press.

Jefferson, J. 1994. Barnwell: a radwaste era ends—but not quite yet. Forum for Applied Research and Public Policy 9(3):88-89.

Johnson, R., and W.L. Petcovic. 1988. What are your chances of communicating effectively with technical or non-technical audiences? Paper presented at the Annual Meeting of the Society for Risk Analysis, Washington, D.C., October 30-November 2.

Kasperson, R.E. 1983. Equity Issues in Radioactive Waste Management. Cambridge: Oelschlager, Gunn and Hain.

Kraus, N., T. Malmfors, and P. Slovic. 1992. Intuitive toxicology: expert and lay judgments of chemical risks. Comments on Toxicology 4(6):441-484.

Krauss, C. 1989. Community struggles and the shaping of democratic consciousness. Sociological Forum 4(2):227-239.

Kunreuther, H., K. Fitzgerald, and T.D. Aarts. 1993. Siting noxious facilities: a test of the facility siting credo. Risk Analysis 13(3):301-318.

League of Women Voters. 1993. The Nuclear Waste Primer: A Handbook for Citizens. New York: Lyons & Burford.

Levine, A.G. 1982. Love Canal: Science, Politics, and People. Lexington, Mass.: Lexington Books.

Lipset, S.M., and W. Schneider. 1983. The decline of confidence in American institutions. Political Science Quarterly 98:379-402.

Licensing Requirements for Land Disposal of Radioactive Waste. Code of Federal Regulations, Title 10, Part 61 (10 CFR 61). 1995 edition.

Lovejoy, L.A., Jr. 1994. New Mexico waste plant sits idle amid controversy. Forum for Applied Research and Public Policy 9(3):113-115.

Low-Level Radioactive Waste Forum. 1996. Summary Report: Low-Level Radioactive Waste Management Activities in States and Compacts. A Supplement to LLW Notes 4(1). Washington, D.C.

Low-Level Radioactive Waste Policy Act. 1980 (P.L. 96-573, December 23). United States Statutes at Large 94:3347-3349.

Low-Level Radioactive Waste Policy Amendments Act of 1985. 1986 (P.L. 99-240, Title I, January 15). United States Statutes at Large 99:1842-1859.

Lynn, F. 1986. The interplay of science and values in assessing and regulating environmental risks. Science, Technology and Human Values 11(2):40-50.

Maharik, M., and B. Fischhoff. 1993. Risk knowledge and risk attitudes regarding nuclear energy sources in space. Risk Analysis 13(3):345-353.

Marrett, C.B. 1984. Public concerns about nuclear power and science. Pp. 307-328 in E.A. Rosa and W.R. Freudenburg, eds., Public Reactions to Nuclear Power:

Are There Critical Masses? Boulder, Colo.: American Association for the Advancement of Science/Westview.

McCabe, G.H. 1994. Radioactive waste: a view from abroad. Forum for Applied Research and Public Policy 9(3):81-85.

McCray, J.G. et al., eds. Waste Isolation in the U.S. Vol. 3: Waste Policies and Programs. Tucson: University of Arizona Press.

McKinley, I.G., and C. McCombie. 1994. Switzerland plans to bury nuclear-waste problem. Forum for Applied Research and Public Policy 9(3):116-118.

Miller, B. 1994. High-level waste: view from Nevada. Forum for Applied Research and Public Policy 9(3):103-105.

Monmonier, M. 1994. Spatial resolution, hazardous waste siting, and freedom of information. Statistical Computing and Graphics 5(1):9-11

Monohan, J., and L. Walker. 1985. Social Science in Law: Cases and Materials. Mineola, N.Y.: Foundation Press.

National Research Council. 1984. Social and Economic Aspects of Radioactive Waste Disposal: Considerations for Institutional Management. Washington, D.C.: National Academy Press.

National Research Council. 1990. Ground Water Models: Scientific and Regulatory Applications. Washington, D.C.: National Academy Press.

National Research Council. 1992. Assessment of the U.S. Outer Continental Shelf Environmental Studies Program: III. Social and Economic Studies. Washington, D.C.: National Academy Press.

National Research Council. 1993. Assessment of the U.S. Outer Continental Shelf Environmental Studies Program: IV. Lessons and Opportunities. Washington, D.C.: National Academy Press.

National Research Council. 1994. Environmental Information for Outer Continental Shelf Oil and Gas Decisions in Alaska. Washington, D.C.: National Academy Press.

National Research Council. 1996. Understanding Risk: Informing Decisions in a Democratic Society. Washington, D.C.: National Academy Press.

Nealey, S.M., and J.A. Hebert. 1983. Public attitudes toward radioactive wastes. Pp. 94-111 in C.A. Walker, L.C. Gould, and E.J. Woodhouse, eds. Too Hot to Handle? Social and Policy Issues in the Management of Radioactive Wastes. New Haven, Conn.: Yale University Press.

New York State Assembly. 1993. 1993-1994 regular sessions. A. 7713. An Act to Amend the Environmental Conservation Law and the Public Authorities Law, in Relation to Enacting the Low-Level Radioactive Waste Management Act and Repealing Sections . . . Relating to Radioactive Waste Management (April 28, 1993).

New York State Department of Environmental Conservation. n.d. Freshwater Wetlands Maps and Classification. 6 NYCRR Part 664.

New York State Department of Environmental Conservation, Bureau of Energy and Radiation. 1987. Final Environmental Impact Statement for Promulgation of 6 NYCRR Part 382: Regulations for Low-Level Radioactive Waste Disposal Facilities (Certification of Proposed Sites and Disposal Methods). 2 vols.

New York State Department of Environmental Conservation, Bureau of Radiation. 1988. Final Generic Environmental Impact Statement for Promulgation of 6 NYCRR Part 381: Regulations for Low-Level Radioactive Waste Transporter Permit and Manifest System. 2 vols.

New York State Department of Environmental Conservation, Bureau of Radiation. 1991. Draft Environmental Impact Statement for Promulgation of 6 NYCRR Part 383: Regulations for Low-Level Radioactive Waste Disposal Facilities. 2 vols.

New York State Department of Environmental Conservation, Bureau of Radiation. 1991. Final Generic Environmental Impact Statement for Promulgation of 6 NYCRR, Subpart 383-6: Financial Assurance Requirements for Low-Level Radioactive Waste Disposal Facilities. 2 vols.

New York State Department of Environmental Conservation, Bureau of Radiation. 1993. 6 NYCRR Part 382: Regulation of Low-Level Radioactive Waste (LLRW) Disposal Facilities: Certification of Proposed Sites and Disposal Methods (March 14, 1993).

New York State Department of Environmental Conservation, Bureau of Radiation. 1993. Final Generic Environmental Impact Statement for Promulgation of 6 NYCRR Part 383: Regulations for Low-Level Radioactive Waste Disposal Facilities (Design, Construction, Operation, Closure, Post-Closure, and Institutional Control). 2 vols.

New York State Department of Environmental Conservation, Bureau of Radiation. 1993. Revised Proposed Rulemaking for Low-Level Radioactive Waste Disposal Facilities: Revisions to Proposed 6 NYCRR Part 383 . . . and Revisions to Proposed Amendments to 6 NYCRR 382 and 380.

New York State Department of Environmental Conservation, Bureau of Radiation. 1994. Correspondence to I. Alterman from P. Merges, July 19, regarding response to telephone call of July 14, and transmitting correspondence from U.S. Nuclear Regulatory Commission to Department of Environmental Conservation, June 2, 1992, regarding compatibility of regulations.

New York State Department of Environmental Conservation, Bureau of Radiation. 1994. Correspondence to I. Alterman from N. Nosenchuck, August 19, responding to NAS-NRC committee questions of August 8, with attachments.

New York State Department of Environmental Conservation, Bureau of Radiation. 1994. Correspondence to I. Alterman from N. Nosenchuck, October 21, responding to NAS-NRC committee questions of September 13, with attachments.

New York State Department of Environmental Conservation, Division of Fish and Wildlife. n.d. Freshwater Wetlands Permit Requirements Regulations. 6 NYCRR Part 663 (leaflet).

New York State Department of Environmental Conservation, Division of Fish and Wildlife. 1992 (reprint). Article 24: Freshwater Wetlands, and Title 23 of Article 71 of the Environmental Conservation Law (leaflet).

New York State Department of Health. 1994. Video: ITSEP Meetings: NAS Committee to Review NYS Siting & Methodology Selection for the LLRW Disposal, January 20-21, Albany, N.Y. Tapes 1-3.

New York State Department of Health. 1994. Video: ITSEP Meetings: NAS Committee to Review NYS Siting & Methodology Selection for the LLRW Disposal, March 24-25, Ithaca, N.Y. Tapes 1-2.

New York State Department of Health. 1994. Video: ITSEP Meetings: NAS Committee to Review NYS Siting & Methodology Selection for the LLRW Disposal, August 22, Syracuse, N.Y. Tape 1.

New York State Department of Health. 1994. Video: ITSEP Meetings: NAS Committee to Review NYS Siting & Methodology Selection for the LLRW Disposal, November 1-2, Albany, N.Y. Tapes 1-5.

New York State Energy Research and Development Authority. 1993. New York State Low-Level Radioactive Waste Status Report for 1992.

New York State Energy Research and Development Authority. 1994. New York State Low-Level Radioactive Waste Status Report for 1993.

New York State Energy Research and Development Authority. 1995. New York State Low-Level Radioactive Waste Status Report for 1994.

New York State Legislature. 1986. Low-Level Radioactive Waste Management Act. Laws of the State of New York, Chapter 673 (July 26):2801-2817.

New York State Legislature. 1990. An Act to Amend the Environmental Conservation Law in Relation to the Membership and Duties of the Commission . . . and the Advisory Committee . . . [amendment to 1986 LLRW Management Act]. 1990 Session Laws, Chapter 913 (A. 12080, July 30).

New York State Low-Level Radioactive Waste Siting Commission. 1988. Candidate Area Identification Report. Albany, N.Y.

New York State Low-Level Radioactive Waste Siting Commission. 1988. Generic Scope of Draft Environmental Impact Statement. Albany, N.Y.

New York State Low-Level Radioactive Waste Siting Commission. 1988. Public Meeting Summary Report, October 3-14. Albany, N.Y.

New York State Low-Level Radioactive Waste Siting Commission. 1988. Plan for Selecting Sites for Disposal of Low-Level Radioactive Wastes. Albany, N.Y.

New York State Low-Level Radioactive Waste Siting Commission. 1988. Statewide Exclusionary Screening Report. Albany, N.Y.

New York State Low-Level Radioactive Waste Siting Commission. 1988. Summary of the Site and Method Workshop, Albany, N.Y., August 5-7. Albany, N.Y.

New York State Low-Level Radioactive Waste Siting Commission. 1989. Public Meeting Summary Report, January 18-26. Albany, N.Y.

New York State Low-Level Radioactive Waste Siting Commission. 1989. Disposal Method Screening Report. Albany, N.Y.

New York State Low-Level Radioactive Waste Siting Commission. 1989. Report on Potential Sites Identification; Executive Summary; Appendix A, Allegany; Appendix B, Cayuga; Appendix C, Chenango North; Appendix D, Chenango South; Appendix E, Clinton; Appendix F, Cortland; Appendix G, Montgomery; Appendix H, Orange/Ulster; Appendix I, Oswego; Appendix J, Washington. Albany, N.Y.

New York State Low-Level Radioactive Waste Siting Commission. 1990. Evaluation of the Feasibility of the Mined Repository Option, March.

New York State Low-Level Waste Siting Commission. 1990. Video: Pre-Characterization, State University of New York, Cortland, February 8.

New York State Low-Level Waste Siting Commission. 1991. Video: Site Selection Presentation, Part I, July 17. Tapes 1-2.

New York State Low-Level Waste Siting Commission. 1991. Video: Site Selection Presentation, Part II, September 11. Tapes 1-3.

New York State Low-Level Waste Siting Commission. 1991. Video: Site Selection Presentation, Part III, October 16. Tapes 1-2.

New York State Low-Level Waste Siting Commission. 1991. Video: Site Selection Presentation, Part IV, November 14. Tapes 1-2.

New York State Low-Level Radioactive Waste Siting Commission. 1992. Response to Public Comments Relating to the Report on Potential Sites Identification. Albany, N.Y.

New York State Low-Level Radioactive Waste Siting Commission. 1993. Excluded Areas Report. Albany, N.Y.

New York State Low-Level Radioactive Waste Siting Commission. 1994. Correspondence to I.B. Alterman from D.A. Eldridge, November 17, responding to questions raised during the November 1994 Albany meeting and transmitting supporting documentation on public participation: list of members of first Advisory Committee; pre-1990 Public Participation Plan; Public Outreach Plan and Media Plan for September 1989-January 1990; January 18, 1990, Public Participation Program Status Report.

New York State Low-Level Radioactive Waste Siting Commission. 1994. Correspondence to I.B. Alterman from B. Goodale, November 2, transmitting supporting documentation on questions raised concerning Allegany County: February 20, 1990, Siting Commission response to the Allegany County study on the CAIR.

New York State Low-Level Radioactive Waste Siting Commission. 1994. Correspondence to I. Alterman from J. Williams, August 4, transmitting Glossary and References from 1988 draft "Plan for Selecting Sites for Disposal of Low-Level Radioactive Waste."

New York State Low-Level Waste Siting Commission. 1994. Correspondence to I. Alterman from S. Wilson, December 13, transmitting Plan for Selecting Methods for Disposal of Low-Level Radioactive Wastes, November 1988; and chronological list of commission's major activities and publications.

New York State Low-Level Radioactive Waste Siting Commission. 1994. Correspondence to G. Anderson from S. Wilson, October 11, transmitting supporting documentation on pre-1990 public meetings: LLRW Management Public Opinion Research Report, Gordon S. Black Corp., December 14, 1993; 35 Attachments and three appendices, Public Meeting Summary Report.

New York State Low-Level Radioactive Waste Siting Commission. 1994. Correspondence to W.R. Freudenburg from J.S. Williams, September 16, responding to information request of September 2, concerning date of volunteer policy resolution.

New York State Low-Level Radioactive Waste Siting Commission. 1994. Correspondence to C. Monaco from T. Ostrander, October 20, regarding information request of July 4, and transmitting county comments on "Plan for Selection of Sites for the Disposal of Low-Level Radioactive Wastes," "Plan for Selection of Methods for the Disposal of Low-Level Radioactive Waste," "Disposal Method Screening Report," from Allegany, Cortland, Oswego, and St. Lawrence counties; Commission "Response to Revised ITSEP Questions November 1994"; and information on volunteered sites, windshield surveys; and commission memo regarding "Incompatible Structures Report."

New York State Low-Level Radioactive Waste Siting Commission. 1994. Correspondence to S. Wiltshire from A. Orazio, September 1, regarding incorrect statement in Steve Wilson's testimony before the committee, August 22.

New York State Low-Level Radioactive Waste Siting Commission. 1994. Newsline (newsletter).

New York State Low-Level Radioactive Waste Siting Commission. 1994. New York State Low-Level Radioactive Waste Siting Commission's Response to ITSEP Questions: National Academy of Science/National Research Council Meeting, August 22. Memorandum.

New York State Low-Level Radioactive Waste Siting Commission. 1995. Correspondence to I.B. Alterman from J. Dunkelberger with copy to R. Ahrens, April 25, transmitting requested copies of the GIS printouts from GIS digitized data portraying the class 1-4 agricultural soils for the Cortland and Allegany candidate areas.

New York State Low-Level Radioactive Waste Siting Commission. 1995. Correspondence to I.B. Alterman from J. Dunkelberger with copy to W. Freudenburg, May 2, transmitting requested copies of GIS produced worksheets from population data provided by Census Bureau for Allegany, Cortland, Orange, and Oswego counties. Includes population, area, etc., along with calculated population density.

New York State Low-Level Radioactive Waste Siting Commission. 1995. Correspondence to J. Croes from S. Wilson, with copy to I.B. Alterman, March 8, transmitting (1) Quality Assurance Plan for the New York State LLRW Siting Commission Siting and Method Selection Study, Rev. 1, December 1989, (2) QMP No. 2.1; Assignment of QA Levels and QA Controls, August 1989, (3) New York State LLRW Siting Commission Quality Assurance Plan, December 1989.

New York State Low-Level Radioactive Waste Siting Commission. 1995. Final Report. Albany, N.Y.

New York State Low-Level Radioactive Waste Siting Commission. 1995. 1994 Source Term Report Executive Summary. Albany, N.Y.

New York State Museum/Geological Survey; State Education Department, University of the State of New York. 1990. New York State Geological Highway Map. State Education Department, Albany, New York.

New York State Museum/Geological Survey; State Education Department, University of the State of New York. 1991. Geology of New York: A Simplified Account. State Education Department, Albany, New York.

Parker, F.L. 1994. Outlook remains dim for waste solution. Forum for Applied Research and Public Policy 9(3):98-102.

Pijawka, D.K., and A.H. Mushkatel. 1992. Public opposition to the siting of the high-level nuclear waste repository: the importance of trust. Policy Studies Review 10:180-94.

Rosa, E.A., and W.R. Freudenburg, eds. 1984. Public Reactions to Nuclear Power: Are There Critical Masses? Boulder, Colo.: American Association for the Advancement of Science/Westview.

Sierra Club Atlantic Chapter. 1994. Correspondence to C. Hornibrook from M. Rose, August 31, regarding Sierra Club Atlantic's view on LLRW siting in New York State, with attachments.

Slovic, P. 1987. Perception of risk. Science 236:280-285.

Slovic, P. 1991. Perceived risk, trust, and the politics of nuclear waste. Science 254:1603-1607.

Slovic, P. 1993. Perceived risk, trust, and democracy. Risk Analysis 13(6): 675-682.

Slovic, P., J.H. Flynn, and M. Layman. 1991. Perceived risk, trust, and the politics of nuclear waste. Science 254(December 13):1603-1607.

Sullivan, T.M., and M. Chehata. 1995. Overview of Research and Development in Subsurface Fate and Transport Modeling (BNL-52469). Upton, New York: Brookhaven National Laboratory.

Szymalak, J.N. 1994. Dear Committee on Open Government: why can't you tell us what constitutes a public body subject to the Open Meetings Law? Unpublished manuscript.

Taffet, M.J., A.L. Lamarre, and J.A. Oberdorfer. 1991. Performance of a mixed-waste landfill amid geologic uncertainty—learning from a case study: Altamont Hills, California, U.S.A. Environmental Geology and Water Sciences 18(3):185-194.

U.S. Department of Energy, National Low-Level Waste Management Program. 1993. 1992 State-by-State Assessment of Low-Level Radioactive Wastes Received at Commercial Disposal Sites (DOE/LLW-181). Idaho Falls, Idaho.

U.S. Department of Energy, National Low-Level Waste Management Program. 1994. Comparative Approaches to Siting Low-Level Radioactive Waste Disposal Sites by W.F. Newberry (DOE/LLW-199). Idaho Falls, Idaho.

U.S. Department of Energy, National Low-Level Waste Management Program. 1994. Directions in Low-Level Waste Management: A Brief History of Commercial Low-Level Radioactive Waste Disposal (DOE/LLW-103 Rev. 1).

U.S. General Accounting Office. 1992. Nuclear Waste: New York's Adherence to Site Selection Procedures Is Unclear. Report prepared for the Honorable Alfonse M. D'Amato, U.S. Senate (GAO/RCED-92-172). Washington, D.C.

U.S. Nuclear Regulatory Commission. 1983. Final Technical Position on Documentation of Computer Codes for High-Level Waste Management (NUREG-0856). Washington, D.C.

U.S. Nuclear Regulatory Commission. 1988. Peer Review for High-Level Nuclear Waste Repositories: Generic Technical Position (NUREG 1297). Washington, D.C.

U.S. Nuclear Regulatory Commission. 1988. Qualification of Existing Data for High-Level Nuclear Waste Repositories (NUREG 1298). Washington, D.C.

U.S. Nuclear Regulatory Commission. 1988. Standard Format and Content of a License Application for a Low-Level Radioactive Waste Disposal Facility—Safety Analysis Report. Rev. 1 (NUREG-1199). Washington, D.C.

U.S. Nuclear Regulatory Commission. 1989. Quality Assurance Guidance for Low-Level Radioactive Waste Disposal Facility (NUREG-1293).

U.S. Office of Technology Assessment. 1985. Managing the Nation's Commercial High-Level Radioactive Waste. Washington, D.C.: U.S. Government Printing Office.

U.S. Secretary of Energy Advisory Board, Task Force on Radioactive Waste Management. 1993. Earning Public Trust and Confidence: Requisites for Managing Radioactive Wastes. Washington, D.C.: U.S. Department of Energy.

Visocki, K., and S.S. Bremon. 1994. Regional compacts and waste disposal. Forum for Applied Research and Public Policy 9(3):86-90.

Vyner, H.M. 1988. Invisible Trauma: The Psychosocial Effects of Invisible Environmental Contaminants. Lexington, Mass.: Lexington Books

Walker, C.A., L.C. Gould, and E.J. Woodhouse, eds. 1983. Too Hot to Handle? Social and Policy Issues in the Management of .Radioactive Wastes. New Haven, Conn.: Yale University Press.

Weibel, C.P., and R.C. Berg. 1991. Importance of geologic characterization of potential low-level radioactive waste disposal sites. Environmental Geology and Water Sciences 18(3):209-214.

Wiltshire, S. 1989. Three necessary conditions for progress in low-level waste management: political commitment, managerial skill, and public involvement. Paper presented at Eleventh Annual DOE Low-Level Waste Conference, Pittsburgh, Pa., August 23.

Additional materials reviewed:

Binder of selected responses from New York State counties identified in the *Candidate Area Identification Report*. Compiled by consultant.

Brochures and newspaper articles relating to radiation, siting, and nuclear waste. Provided by D. Gage.

Various newspaper and trade press articles relating to low-level radioactive waste in New York State and the Northeast.

Cortland County Low-Level Waste Office, Documents Submitted to the National Research Council Board on Radioactive Waste Management's Committee to Review New York State's Siting and Methodology Selection for Low-Level Radioactive Waste Disposal, three submissions:

Submission Number 1:

1. LLRW Articles, Volume 1, Numbers 1-24, 3/28/90-1/3/91
2. LLRW Storage Issues: Cortland's Perspective, 11/19/91-11/21/91
3. Cortland County LLRW Office Special LLRW Bulletin, 7/31/91
4. Cortland County LLRW Office Newsletter No. 1, 4/91
5. Cortland County LLRW Office Newsletter No. 2, 5/91
6. Cortland County LLRW Office Newsletter No. 3, 6/91
7. Cortland County LLRW Office Newsletter Nos. 4 and 5, 7/91-8/91
8. Cortland County LLRW Office Newsletter Nos. 5 and 6, 9/91-10/91
9. Cortland County LLRW Office Newsletter Nos. 7 and 8, 11/91-12/91
10. Implications of Psychological Research on Stress and Technological Accidents. American Psychologist, 6/93
11. Taylor Against LLRW: Presentation on the Issue of Incompatible Structures, 11/21/91
12. Excluded Areas Report: Johnson to Orazio, 8/10/92
13. Comments on Excluded Areas Report: Cote-Hopkins to CAC, 8/1/92
14. The Siting Commission Evaluation of Geology and Mines, 11/14/92
15. Evaluation of Geology and Mines, Follow-up Correspondence, Siting Commission to Snyder, 4/17/92
16. Evaluation of Geology and Mines, Follow-up Correspondence, Snyder to Siting Commission, 5/28/92
17. GIS Correspondence: Cortland County to Siting Commission, 5/2/91
18. GIS Correspondence: Cortland County to Siting Commission, 4/8/91
19. GIS Correspondence: Siting Commission to Cortland County, 3/25/91
20. GIS Correspondence: Cortland County to Siting Commission, 1/23/91
21. Public Participation/Administration Issues (File Composite):
    - Cortland County to Semick, 8/15/91
    - Press Release: Cortland County Planning Department, 1/15/90
    - Articles from *The Cortland Standard*:
        - County Comes Step Closer to Getting Siting Panel Report, 1/24/90
        - N-Dump Lawyer: No Facts in Report, 1/16/90
        - County Argues in Court to Obtain Siting Report, 1/17/90

Dump Panel Defends Position in Report, 1/19/90
Siting Commission Defends Stance to Deny Public Access in Its Study, 1/22/90
Judge Ok's New Version of Report, 1/24/90
UUP Opposes N-Dump, 2/12/90
Professors' Union Calls for Orazio to Resign, 2/5/90
Remarks Insulting and Inflammatory, 1/26/90
Press Release from Cindy Monaco, 2/15/90
Judge Orders Report on N-Dump Released, 2/16/90
County to Get Crucial Dump Report in Entirety, 2/16/90
One Man's Experience, 1/19/90
Invitation Is Declined, 2/13/90
$200/Day to Doodle, 2/8/90
Providing a Forum, 2/13/90
Cuomo Budget Triples Nuclear Waste Dump Panel Funding, 1/26/90
Feuss: County Needs Info Before It'll Meet Commission, 1/26/90
Siting Panel Seeks Talks to Avoid Future Clashes, 1/25/90
State Allowed on Taylor Land Pending Appeal, 2/1/90
State Nuke Dump Panel Admits Credibility Gap, 2/1/90
Dump Test Document May Be Released to County, 2/1/90
N-Dump Panel, County to Meet, 2/7/90
Siting Team to Meet with County Officials Here, 2/7/90
Siting Commission Presents Plans, 2/11/90
Siting Process Is Not Legitimate, Undated (89-90)
Mushrooms Unite, Undated (89-90)
Both Sites Unsuitable, Undated (89-90)
County, Siting Panel Share Data, 2/9/90
County Shown N-Dump Reports, 2/8/90
The Charade Continues, 2/9/90
N-Dump Meeting Disappoints County, 2/10/90
"Deeply Disturbed" by Feb. 9 Editorial, 3/5/90
County Officials: Editorial Was an Accurate Portrayal of Feb. 8th Meeting with Commission, 3/5/90
Reasons for First Dump Tests Changed, 2/27/90
Mr. Orazio Is Confused, 3/5/90
Protesters "Hold Off" at Reporter, 2/7/90
Weston Agrees to Fine in Fraud Case, 2/7/90
N-Dump Consultant Settles EPA Fraud Suit, 2/9/90

22. Technical Issues—Composite File:
    Correspondence:

Cortland County to Murray, 4/10/90
Cortland County to Merges, 4/17/90
Cortland County to Merges, 1/2/90
DOH [New York State Department of Health] to Snyder, 1/11/90
Meeting Minutes: Consideration of Mines in Future Project Activities, 3/17/89
Correspondence: Weston to Goodale, 5/1/89
Preliminary Comments on "Report on Potential Site Identification," Michalski for Cortland County, 9/89
Jonathan Harrington: NYSLLRWSC's Selection of Candidate Areas in Taylor-Harrington for Cortland County, 11/15/89
Correspondence:
USGS to Jarvis, 2/13/90
Sun Exploration & Production Co. to Jarvis & Johnson, 2/28/89
Michalski to Monaco, 1/3/90
Harrington to Monaco, 2/21/90
Michalski to Monaco, 2/21/90
Wesley/Ostmand to Monaco, 2/20/90

23. Miscellaneous Issues:
Correspondence:
Tupper to Heller, 2/23/90
Tupper to Heller, 1/10/90
Heller to Tupper, 1/24/90
Feuss to Dunkelberger, 10/10/89
Feuss to Dunkelberger, 10/2/89
Presentation by Cindy Monaco on behalf of Cortland County, 9/14/89
Correspondence:
O'Mara to Orazio, 9/12/89
Dunkelberger to O'Mara, 2/12/91
DEC to Snyder, 11/20/89
Snyder to Schwartz, 10/31/89
Snyder to Eldridge, 10/31/89
Siting Commission to Snyder, 9/6/89
Cortland County to Eldridge, 4/16/90
Taylor Town Board to Orazio, 2/12/91
LLRW Opposition Endorsement List, 5/23/90
Cortland County to Goodale, 2/7/90
Siting Commission to Monaco, 1/29/90
Cortland County to Goodale, 1/9/90
Letter to the Editor: Pattern of Deceit Seen, 9/17/90

Correspondence:
Siting Commission to Feuss, 10/24/89
Rivest/SUNY Cortland to Monaco, 11/3/89
Hubbard/SUNY Cortland to Monaco, 11/9/89
Tompkins Cortland Community College to Monaco, 11/14/89
SUNY Cortland Tri Beta Statement on the Process of Siting a Radioactive Waste Facility, 5/20/90
Applegate to Cote-Hopkins, 10/31/89
An Informational Community Forum on the LLRW Dump, 11/13/89
Correspondence:
Hubbard/SUNY Cortland to Monaco, 4/13/89
Hubbard/SUNY Cortland to Monaco, 5/5/89
WSKG to Monaco, 10/2/89
WNYP to Monaco, 10/10/89
Cornell University to Monaco, 5/16/90

24. Comment on Draft for Selecting a Preferred Method for Disposal of LLRW: Cortland County to Orazio, 6/7/93
25. USGAO: New York's Adherence to Site Selection Procedures Is Unclear, 8/92
26. Cortland County's Presentation to CAC: Economic Viability of a LLRW Disposal Facility, 9/22/93
27. Cortland County's Testimony Presented to the New York State Legislative Commission on Rural Resources, 10/1/92
28. Cortland County's Commentary on the New York State LLRW Siting Commission's Draft Strategic Plan for Public Participation for Method Selection, 10/20/92
29. Memorandum: Assemblyman Luster to All Senators and Members of Assembly, 12/29/93
30. Correspondence: Kaye to Luster, 12/27/93
31. Editorial: Money Is the Answer, 12/28/93

Submission Number 2:

Binder No. 32: Cortland County's Comments and Considerations Regarding the LLRW Disposal Facility Siting Process, by Cindy Monaco

1. Executive Summary, 11/15/89
2. Section 1: Comments Regarding the CAIR, 11/15/89

3. Section 1, Appendix A: Comments and Considerations Regarding the LLRW Siting Commission's CAIR, by Johnson and Jarvis, 11/15/89
4. Section 1, Appendix B: Questions from Cortland County presented at 21 March 1989 Technical Presentation
5. Section 2: Comments on the Plan for Selecting Sites and Methods for Disposal of LLRW, 11/15/89
6. Section 3: General Observations with regard to the review of the Siting Commission Meeting Minutes, 11/15/89
7. Section 4: Siting Commission Responsiveness, 11/15/89
8. Section 4, Appendix A: 4 April 1989 letter from Cortland County Planning Department to New York State LLRW Siting Commission
9. Section 4, Appendix B: 26 October 1989 letter to Dr. H. David Maillie and Health Questions
10. Section 4, Appendix C: 1 November 1989 letter to Dr. Mortimer Heller from Cortland County Planning Department to New York State LLRW Siting Commission
11. Section 4, Appendix D: 23 October 1989 letter to Jay Dunkelberger from James O'Mara
12. Section 5: Comments Regarding the Report on Potential Sites Identification [RPSI], 11/15/89
13. RPSI: Summary of Previous Screening Steps, 11/15/89
14. RPSI: Overview of the Potential Sites Identification Process, 11/15/89
15. RPSI: Summary of GIS Screening of Candidate Areas, 11/15/89
16. RPSI: Limited Site Inspection, 11/15/89
17. RPSI: Information Received from Local Governments, 11/15/89
18. RPSI: Evaluation of Site-Specific Criteria, 11/15/89
19. RPSI: Comparative Assessments of Potential Sites, 11/15/89
20. Section 5, Appendix A: 31 October 1989 letter to Jay Dunkelberger from Cortland County Planning Department
21. Section 5, Appendix B: 18 October 1989 letter to Bruce Goodale from Cortland County Planning Department
22. Section 5, Appendix C: Status of Agriculture and Agricultural Districts on Identified LLRW Depository Sites in Cortland County, 10/10/89
23. Section 5, Appendix D: Specialty Geologic Consultant Sheets, 5/16/89
24. Section 5, Appendix E: Information from Sun Oil, 2/28/89
25. Section 5, Appendix F: 22 September 1989 letter to Cortland County Planning Department from Bruce Goodale
26. Section 6: Aid to Local Government

27. Section 6, Appendix A: Comments and Considerations with regard to the Host Area Benefits Package, by Schoeberl, Venet, Temple, Petrus, Cieri, and Monaco
28. Appendix A: Fannie Mae Environmental Hazards Mortgage Eligibility Requirements and Underwriting Guidelines, 7/5/89
29. Statement by Ulster County LLRW Task Force, by Schoeberl
30. Appendix B: Considerations for a Health Profile, by Cieri
31. New York State LLRW Stream Analysis, by Hameister, 7/8/89

Submission Number 3:

Binder No. 33: New York State DOH LLRW Information Program Examples and Public Responses

1. DOH Letter to Citizens from Steve Gavitt and Radiation and Health Brochure, 8/89
2. Answers to Questions About Radiation and LLRW, 1/90
3. Editorial: Deception by Omission, 90
4. Letter to Rimawi from Monaco, 7/16/90
5. LLRW Report—Radiation and Health, by Monaco, 7/24/90
6. LLRW Report—Answers to Questions About Radiation and LLRW, by Monaco and Snyder, 7/25/90
7. Letter to Rimawi from Coalition on West Valley Waste, by Vaughan, 8/90
8. Letter to Gavitt from Monaco, 11/5/90
9. LLRW Report—Commercial Military Waste, by Monaco, 11/5/90
10. Memo to Cote-Hopkins from Feuss—Draft Proposal by New York State DOH, 11/7/90
11. Letter to Gavitt from CARD, by Weiss, 11/15/90
12. Letter to Monaco from Gavitt (DOH), Radiation and Health Brochure, 11/21/90
13. Letter to Rimawi from Don't Waste New York, by Weiss, 11/14/90
14. Letter to Gavitt from Cote-Hopkins, 11/16/92
15. Letter to Cote-Hopkins from Gavitt and LLRW Draft Documents—Fact Sheet and LLRW Disposal Overview, 9/24/92
16. Final Draft of LLRW Fact Sheet and LLRW Disposal Overview, 10/93

Appendix F

The Committee's Questions to New York State

**NATIONAL ACADEMY OF SCIENCES-
NATIONAL RESEARCH COUNCIL**

**Committee to Review New York
State's Siting and Methodology Selection for
Low-Level Radioactive Waste Disposal**

Questions for the Siting Commission and/or
the Department of Environmental Conservation
August 22-23, 1994

I. REGULATORY ISSUES

A. The Siting Plan

1. What reviews were done by the Siting Commission (SC) to determine that the Siting Plan meets the intent of the regulations?

2. What role and position did the DEC [New York State Department of Environmental Conservation] take in the need to review and approve that the siting plan meets the regulation? Did SC ask for a formal interpretation by the DEC? Why didn't the DEC take a more pro-active role?

3. Did the SC deviate from the siting plan?

a. If yes, for which parts of the process? Are these deviations documented in any of the reports? If so, where?

b. Did this include treatment of the volunteer (or offered) site at Taylor North? Inasmuch as the site contained mineral soils groups 1-4 (an exclusionary criterion at the site screening stage, Step 3), had a score below the cutoff, and did not have the community's approval, why did the commission maintain this as one of the sites?

4. Was the level of detail in the Siting Plan sufficient to guide adequately the selection of the five potential sites in Step 3?

a. Were exclusionary variables used as preference factors later in the process? If so, when? Where is this documented?

b. According to the site selection plan, preliminary performance assessment analyses would be done before selecting potential sites. It was stated specifically that a preliminary PA would be done for volunteered sites prior to selection of potential sites and reconnaissance evaluation. Were PAs done for the five potential sites? Where is this documented?

B. Criteria

1. Were all the exclusionary criteria derived from the regulations? Were factors used in the site selection process that are not specified in either the regulations or the site selection plan? If so, which were they, when were they applied, and why?

2. What was the basis for:

a. the existing mine exclusion?
b. the nonattainment exclusion?
c. the depth of geologic unit exclusion?

3. Are all the exclusionary criteria included in the Siting Plan?

4. Were the criteria for the "windshield surveys" documented anywhere? What guidelines were used for conducting them to assure consistency from site to site?

5. Who developed the concept of a "fatal flaw" criterion?

a. How is "fatal flaw" defined? Where in the site selection documents is this defined?

b. What criteria were used to determine fatal flaws? What was the basis for the criteria? Where is this found in the site selection record?

c. What authority did the DEC have to grant a variance if a fatal flaw existed?

d. How well did the public understand this?

C. Application of Criteria

1. Exclusionary Criteria

a. In the Step 1 exclusion process, 5 requirements are established for use (page 4-1, Excluded Lands Report). Were other factors like cost or ease of application considered?

b. What are the differences in criteria for above/belowground disposal sites and for mined disposal sites?

c. To what extent were the nonperformance-based exclusionary criteria used to reduce the number of potential sites (e.g., the criteria for nonattainment of air quality standards)?

2. Preference Criteria

a. Could a possible misinterpretation of 10 CFR 61 (61.50(a.5)) have steered the selection toward agricultural soils? The idea of the cited section of 10 CFR 61 (near-surface disposal method) was to allow fairly rapid downward movement of water through the soils to minimize contact time with the waste. This criterion seems to have been interpreted by the Siting Commission to mean good surface drainage away from the site. Is it possible that this difference in interpretation could have led to more runoff-prone clay-rich soils (i.e., agricultural soils) gaining preference over nonclay soils?

b. Were the incremental distances or values used in avoidance criteria technically based (e.g. dose, other rationale)?

c. Was stratigraphic complexity avoided because of the difficulty in modeling it?

d. In the ROPSI [*Report on Potential Sites Identification*], the windshield survey seemed to identify certain criteria as "exclusionary," but actually, as the A, B, and C groupings were used, they are being applied as "preference" criteria. For instance, in the ROPSI (p. 4-40), A, B, and C ratings are based on the number of positive ratings. No PA [performance assessment] analyses appear to have been done to see if the negatives present would outweigh the positives. Nor does it appear that the effects of the negatives on meeting performance objectives were evaluated. This seems to be the case in the comparison of the above/belowground versus drift mines. Please explain. (See page 4-39 ROPSI.)

e. Why were the A, B, C ratings not applied to the volunteered (or offered) sites? (Table 4-7, ROPSI)

D. Quality Assurance Program

1. What was the scope of the Quality Assurance [QA] Program?

2. Who was responsible for implementation?

3. Were there any independent reviews of the QA program? Where are they documented?

E. Process and Procedures

1. Why did the SC decide to complete site selection prior to selecting a disposal method, which they are now required to select first? Is this decision documented?

2. Why didn't the SC do a full statewide exclusionary screening first? Apparently some of the exclusionary criteria were mapped statewide. Is this decision documented?

3. What does the term "desiting" mean to the SC?

4. Why did the SC choose the Intergraph GIS [Geographic Information System] system that was a Weston proprietary system? Were other systems considered? How did they plan to deal with the proprietary aspects of this problem?

5. At which step did the change occur from using 1-mile grids to 40-acre grids for screening? How much area may have been lost in this transfer? Was the public aware of this process?

6. What was the process to determine at what stage the various exclusionary and preference criteria would be applied? How was the public involved in this process? Were any changes made to when the different criteria were applied? If so, what was the justification? Where is this explained in the record?

7. Although not specifically described in the site selection plan, sensitivity analyses of the impact of selected exclusionary criteria upon

identification of potential sites were summarized in the ROPSI (Section 4). Where are the details of these analyses documented? What was the rationale for selection of the factors to be analyzed?

F. Public Involvement Program

1. What recommendations did the SC's pre-1990 citizen advisory committee make to the SC? How did the committee affect decisions made by the SC?

2. What public input did the SC use in making decisions on the siting process and plan, siting criteria, weighting, etc.? Were any changes made as the result of public input in any area of the commission's plans, decisions, process, etc.? If so, please cite.

3. What direction did the SC give staff regarding public involvement? How much time and attention were given to public involvement relative to the entire project? How were resources (time and money) allocated for the public involvement process?

II. TECHNICAL ISSUES

A. General

1. Although the Groundwater Hydrology siting factor uses the terms "primary public water supply aquifer", and "principal aquifer," there is no definition of the term "aquifer."

a. How is the term "aquifer" defined in the Siting Commission's usage: by permeability, degree of saturation, extent, etc.?

b. What are the boundaries of the aquifer? How can the boundaries be determined without subsurface information?

2. Did the Siting Commission have discretion to use products and maps produced by other agencies beside DEC and DOH [New York State Department of Health]? How was it decided which maps of aquifers were to be used such that, based on its extent, the aquifer would be considered exclusionary rather than preferential?

3. As criteria were added with each phase of the site screening process, were the new criteria uniformly applied?

4. As we understand it, certain criteria are exclusionary and constitute a fatal flaw. Do these same criteria become weighting factors later in the process, or are they always fatal flaws regardless of evaluation stage?

5. If new information becomes available that is exclusionary with respect to a specific area or site, is it treated the same way as if it had been known from the beginning of the process?

6. Why were active mines and other areas of mineral resource activity, such as oil and gas recovery or exploration, salt, etc., not exclusionary but were given only preferential status? Did the recent salt mine flooding result in reconsideration of preferred distances to these resource activities?

7. Given the difficulties and extra work involved in screening and trying to select a site that would satisfy both above/belowground and mined facilities, why did the SC not consider selecting the disposal method before continuing to select a site? In selecting candidate areas and sites, why were potential mine areas/sites not selected separately from near-surface facility areas/sites?

8. Why did the SC average the weighting factors for aboveground and belowground facilities?

a. Would the results of weighting the above and belowground methods separately have changed the map of excluded areas?

b. When the method is selected, will the original average or mean weighting factors be applied in the siting process, or will they develop new weighting factors?

9. Why did the SC not distinguish between drift mines and deep mines in weighting and scaling values, but rather used the weights and scales for drift mines that were developed for deep mines?

B. Depth of Mines

1. Although the DEC defines "geologic unit" in the glossary accompanying Part 382 as "the geologic media in which an underground mined

repository is constructed," some aspects of the siting regulation and the criteria with respect to mines are unclear. What is the rationale for the 30-m-depth minimum?

2. As depth to a geologic unit can be spatially variable over very short distances, where could reliable statewide data for all areas of the state be obtained?

a. How was screening done for the 30-m depth?

b. Was an area eliminated if somewhere within its borders bedrock came within 30 m of the surface? How would this be determined?

3. In the siting plan, criterion 4 for existing and new mine methods requires excluding "all abandoned mines and *all geologic units that are less than or equal to 30 meters below the ground surface*" (emphasis added). There are at least two interpretations: either eliminate mines less than 30 m from the surface, or eliminate bedrock that has some portion within 30 m of the surface. How did the SC interpret it?

4. The above quotation is apparently a restatement of the regulation with the term "geologic unit" added. Where did this concept originate? Is this a Siting Commission addition and, if so, should it be a preference criterion?

5. Does the rock type (geologic unit composition, e.g., limestone, granite, shale, etc.) play any role in the criterion?

C. Ground Water Aquifers

1. Although the Part 382 glossary defines "primary public water supply aquifer" and "principal aquifer" in nonquantitative terms, there are quantitative implications. In this context, how do you determine what is "highly productive," "substantial recharge," "potentially abundant source of water," without quantitative guidelines?

2. While we recognize that the aquifers considered in the regulations are unconsolidated nonbedrock deposits, when the criteria were applied later did they include depth limits, clear boundaries, appropriate scale, etc.? Why did DEC exclude bedrock aquifers from consideration in the regulations?

3. Are there aquifer boundaries that cross geologic or lithologic boundaries? In other words, are they mapping alluvium and not aquifers?

4. If you have a municipality using an alluvial aquifer in one small area, does it exclude all areas underlain by the alluvial formation? How would the exclusionary boundary be established? What size does a municipality have to be to determine that the aquifer is a primary or principal aquifer?

5. Two ground water hydrology preferential criteria concern distances: criterion 12 refers to preferential distance from primary or principal aquifers, and criterion 13 considers distance from ground water discharge zones. Terms such as "adequate distance from" and "sufficiently long pathway" are used in the regulation and discussion but no quantitative measure is provided as guidance. What calculations or considerations led to the scaling of the distances of 1/2 mile, greater than 1/2 to 1, and greater than 1? On what basis were these distances determined to be "adequate" or "sufficient"?

6. Is all of Long Island underlain by the Long Island aquifer? If not, what parts of Long Island have a water supply different from that aquifer? What are the boundaries of the Long Island aquifer?

D. Wetlands

1. The Siting Plan makes reference to an executive order of 1977 and ECL [Environmental Conservation Law] Article 24 for the definition of freshwater wetlands. Since wetland definitions have been changing at the federal and state levels, please define wetlands as they have been used by the Siting Commission in the site screening process. What quantitative considerations were used, e.g., areal extent, duration of wet condition, etc.?

2. Although wetlands are excluded pursuant to several cited laws and regulations, the exclusionary criterion 17 and preference criterion 18 (distance from wetlands) are applied only at Step 3 during potential site identification. In the Excluded Areas Report the rationale for applying these criteria during Step 3 rather than at an earlier step is that the wetlands are generally of small size. Please explain then the application of preference criterion 22 (location of facility relative to surface waters) during Step 2, if the wetlands areas are too small to identify during that stage of the screening process. Why are other wetlands preference criteria used only for comparison and not for screening? How do they differ in impact from criterion 22?

E. Air Quality Nonattainment Areas

1. Although DEC did not specifically include air quality in Part 382, the SC, following EPA [U.S. Environmental Protection Agency] guidelines for air quality standards, established exclusionary criterion 26. This exclusion was applied at Step 2, even earlier than wetlands, which is mandated by Part 382. Considering that, as stated in the EAR, "a low level radioactive waste facility is not expected to be a major source of air pollutants," except for "some fugitive dust" during construction, please explain why this criterion was:

a. considered necessary by the Siting Commission;
b. exclusionary, rather than preferential;
c. applied at Step 2, such that it had the effect of excluding an entire county (Nassau) that may not have been excluded for any other reason.

2. What data were used to determine that all of Nassau County did not meet EPA air quality standards?

F. Population Density

1. How was the population density calculation done? What data were used to determine population and areal extent that resulted in the exclusion of specific population centers and the inclusion of other areas?

2. How were the preferential criteria for population density intervals and scale selected?

G. Mineral Soil Groups 1-4

1. What calculations were done to determine the centroids?

2. Why did DEC include soil groups 3 and 4 in the regulations?

3. Why did DEC include regulations related to farms?

NATIONAL ACADEMY OF SCIENCES-NATIONAL RESEARCH COUNCIL
Committee to Review New York State's Siting and Methodology Selection for Low-Level Radioactive Waste Disposal

Questions for the Siting Commission and/or the Department of Environmental Conservation
September 1994

I. REGULATORY ISSUES

A. The Siting Plan

1. Is the A, B, and C rating system in ROPSI (p. 4-40) properly documented in the Siting Plan?

2. Are the criteria for the windshield surveys documented anywhere? What guidelines were used for conducting them to assure consistency from site to site?

3. When was the incineration option, which apparently was the basis for the air attainment criterion, first included and discussed in the Siting Plan?

B. Criteria

1. Were all the exclusionary criteria derived from the regulations? Were criteria and procedures used in the site selection process that are not specified in either the regulations or the site selection plan? If so, where are they documented and what are their bases? [Examples—windshield survey criteria, nonattainment exclusion]

2. Who developed the concept of fatal flaw criterion? (See Site Selection Plan, p. 2-3, Step 4.)

a. How is fatal flaw defined? Where in the siting documents is this defined?

b. What criteria were used to determine fatal flaws? What was the basis for the criteria? Where is this found in the site selection record?

c. How well did the public understand this?

C. Application of Criteria

1. Exclusionary Criteria

a. In the Step 1 exclusion process, five requirements are established for use (p. 4-1, Excluded Lands Report). Were other factors like cost or ease of application considered?

b. What are the differences in criteria for above/belowground disposal sites and for mined disposal sites?

c. To what extent were the nonperformance-based exclusionary criteria used to reduce the number of potential sites (e.g., the criteria for nonattainment of air quality standards)?

2. Preference Criteria

a. Could a possible misinterpretation of 10 CFR 61 (61.50(a.5)) have steered the selection toward agricultural soils? The idea of the cited section of 10 CFR 61 (near-surface disposal method) was to allow fairly rapid downward movement of water through the soils to minimize contact time with the waste. This criterion seems to have been interpreted by the Siting Commission to mean good surface drainage away from the site. Is it possible that this difference in interpretation could have led to more runoff-prone clay-rich soils (i.e., agricultural soils) gaining preference over nonclay soils?

b. Were the incremental distances or values used in avoidance criteria technically based (e.g., dose, other rationale)?

c. Was stratigraphic complexity avoided because of the difficulty in modeling it?

d. In the ROPSI the windshield survey seemed to identify certain criteria as "exclusionary," but actually, as the A, B, and C groupings were used, they are being applied as "preference" criteria. For instance, in the ROPSI (p.4-40), A, B, and C ratings are based on the number of positive ratings. No PA analyses appear to have been done to see if the negatives present would outweigh the positives. Nor does it appear that the effects of the negatives on meeting performance objectives were evaluated. This seems to be the case in the comparison of the above/belowground versus drift mines. Please explain. (See p. 4-39, ROPSI.).

e. Why did drift mines consistently score higher? Provide examples of sensitivity studies conducted at this phase.

f. Why were the A, B, and C ratings not applied to the volunteered or offered sites? Explain how these ratings were used in the process? (Table 4-7, ROPSI).

D. Quality Assurance Program

1. How were the up-front planning and regulatory requirements identified and addressed before the actual work began?

2. How were the quality assurance (QA) procedures (or other procedures) prioritized to ensure that early work would be documented and controlled by a systematic approach?

3. How were the early siting requirements identified and how were personnel trained and qualified to document the activities that would stand up in later challenges by the regulators, among others.?

4. Describe present ownership of the quality program and how this ownership has changed since the program was first developed.

E. Process and Procedures

1. Why didn't the SC do a full statewide exclusionary screening first? Apparently, some of the exclusionary criteria were mapped statewide. Is this decision documented?

2. What does the term "desiting" mean to the SC? Why and how did the SC decide to leave the sites in "limbo" (still under consideration)?

3. Why did the SC choose the Intergraph GIS system that was a Weston proprietary system? Were other systems considered? How did they plan to deal with the proprietary aspects of this problem?

4. At which step did the change occur from using 1-mile grids to 40-acre grids for screening? How much area may have been lost or gained in this transfer? Was the public aware of this process?

5. What was the process to determine at what stage the various exclusionary and preference criteria would be applied? How was the public involved in this process? Were any changes made to when the different criteria

were applied? If so, what was the justification? Where is this explained in the record? Focus on the application aspects.

6. Although not specifically described in the site selection plan, sensitivity analyses of the impact of selected exclusionary and preference criteria upon identification of potential sites were summarized in the ROPSI (Section 4). Where are the details of these analyses documented? What was the rationale for selection of the factors to be analyzed? See question on A, B, and C criteria in Section C 2.d.

7. What documentation establishes the basis for GIS data sets?

a. What documentation establishes the application of these data sets?
b. Is the GIS documentation adequate to establish that the best-available data has been used?
c. Is there reasonable assurance that the results are reproducible?

F. Public Involvement Program

1. What recommendations did the SC's pre-1990 Citizen Advisory Committee (CAC) make to the SC? How did the committee affect decisions made by the SC? How effective were CAC interactions with the SC? What were the difficulties under which the CAC operated?

2. What public input did the SC use in making decisions on the siting process and plan, siting criteria, weighting, etc.? Were any changes made as the result of public input in any area of the commission's plans, decisions, process, etc.? If so, please cite.

3. What direction did the SC give staff regarding public involvement? How much time and attention were given to public involvement relative to the entire project? How were resources (time and money) allocated for the public involvement process?

4. What steps did you take to assure that public values could have an effect on the weighting of preference criteria? What were the factors that limited and/or facilitated your ability to include their input?

5. DOH: How did DOH carry out its responsibility to inform the public about low-level waste?

6. DOH: Have all stakeholders been identified and kept informed (minimally) and involved in the process? If not, why?

7. Citizen groups have said that they received no information on the Siting Commission's decisions of January 1989 on volunteered/offered sites until much later. When were these decisions first made public, according to your records? What documentation, if any (news reports, letters of response), can you provide to this committee to clarify this discrepancy? (Note that citizens have claimed that no minutes from this meeting were made available until much later, so it would be valuable if this answer could go beyond the meeting minutes themselves.)

8. Citizens in Cortland County expressed the view that their technical input was generally ignored—e.g., comments in the period between CAIR [*Candidate Area Identification Report*] and ROPSI that should have changed the favorability scores for specific Cortland County sites. Did the ROPSI fail to respond to this technical input? If so, why? If not, what changes were made in response to technical input from affected regions?

II. TECHNICAL ISSUES

A. General

1. As criteria were added with each phase of the site screening process, were the new criteria consistently applied?

2. As we understand it, certain criteria are exclusionary and constitute a fatal flaw. If so, which? Do these same criteria become weighting factors later in the process, or are they always fatal flaws regardless of evaluation stage?

3. If new information becomes available at various stages that is exclusionary with respect to a specific area or site, is it treated the same way as if it had been known from the beginning of the process?

4. Why were active mines and other areas of mineral resource activity, such as oil and gas recovery or exploration, salt, etc., not exclusionary but were given only preferential status? Did the recent salt mine flooding result in reconsideration of preferred distances to these resource activities?

5. Given the difficulties and extra work involved in screening and trying to select a site that would satisfy both above/belowground and mined facilities, why did the SC not consider selecting the disposal method before continuing to select a site?

6. In selecting candidate areas and sites, why were potential mine areas/sites not selected separately from near-surface facility areas/sites? At what stage in the site selection process does consideration of mining technologies become involved?

7. Why did the SC average the weighting factors for aboveground and belowground facilities?

a. Would the results of weighting the surface and subsurface mining methods separately have changed the map of excluded areas?

b. When the method is selected, will the original average or mean weighting factors be applied in the siting process, or will they develop new weighting factors?

c. Why did the SC not distinguish between drift mines and deep mines in weighting and scaling values?

B. Ground Water/Aquifers

SC: 1. Although the Part 382 glossary defines "primary public water supply aquifer" and "principal aquifer" in nonquantitative terms, there are quantitative implications. In this context, how do you determine what is "highly productive," "substantial recharge," "potentially abundant source of water," without quantitative guidelines (volume, thickness, areal extent of aquifers, nature of confining units)?

SC: 2. Are there aquifer boundaries that cross geologic or lithologic boundaries? How are they delineated?

SC: 3. If a municipality uses an alluvial aquifer in one small area, are all areas underlain by the alluvial formation also excluded? How would the exclusionary boundary be established? What size does a municipality have to be to determine that the aquifer is a primary or principal aquifer?

DEC and SC: 4. Two groundwater hydrology preferential criteria concern distances: criterion 12 refers to preferential distance from primary or principal aquifers, and criterion 13 considers distance from groundwater discharge zones. Terms such as "adequate distance from" and "sufficiently long pathway" are used in the regulation and discussion, but no quantitative measure is provided as guidance. What calculations or considerations led to the scaling of the distances of 1/2 mile, greater than 1/2 to 1, and greater than 1? On what basis were these distances determined to be "adequate" or "sufficient?"

C. Wetlands

SC and DEC: 1. The Siting Plan makes reference to an executive order of 1977 and ECL Article 24 for the definition of freshwater wetlands. Since wetland definitions have been changing at the federal and state levels, please define wetlands as they have been used by the Siting Commission in the site screening process. What quantitative considerations were used, e.g., areal extent, duration of wet condition, etc.?

SC and DEC: 2. Although wetlands are excluded pursuant to several cited laws and regulations, the exclusionary criterion 17 and preference criterion 18 (distance from wetlands) are applied only at Step 3 during potential site identification. In the Excluded Areas Report the rationale for applying these criteria during Step 3 rather than at an earlier step is that the wetlands are generally of small size. Please explain then the application of preference criterion 22 (location of facility relative to surface waters) during Step 2, if the wetlands areas are too small to identify during that stage of the screening process. Why are other wetlands preference criteria used only for comparison and not for screening? How do they differ in impact from criterion 22?

D. Other Criteria

1. Air Quality Nonattainment Areas: What data were used to determine that all of Nassau County did not meet EPA air quality standards?

2. Population Density: In choosing the scales and weights for population density criteria, did SC take into account the effect of the proposed site on future population density?

3. Mineral Soil Groups 1-4: How were the centroids of the map unit delineations determined? Could these data be reproduced? If not, was any sensitivity analysis performed for replication?

Appendix G

Classification of Wastes

(From the Code of Federal Regulations)[1]

Title 10—Energy
Chapter I—Nuclear Regulatory Commission
Part 61—Licensing Requirements for Land Disposal of Radioactive Waste
Subpart D—Technical Requirements for Land Disposal Facilities

§ 61.55 Waste classification.

(a)…(2) Classes of waste.

(i) Class A waste is waste that is usually segregated from other waste classes at the disposal site. The physical form and characteristics of Class A waste must meet the minimum requirements set forth in § 61.56(a). If Class A waste also meets the stability requirements set forth in § 61.56(b), it is not necessary to segregate the waste for disposal.

(ii) Class B waste is waste that must meet more rigorous requirements on waste form to ensure stability after disposal. The physical form and characteristics of Class B waste must meet both the minimum and stability requirements set forth in § 61.56.

(iii) Class C waste is waste that not only must meet more rigorous requirements on waste form to ensure stability but also requires additional measures at the disposal facility to protect against inadvertent intrusion. The physical form and characteristics of Class C waste must meet both the minimum and stability requirements set forth in § 61.56.

(iv) Waste that is not generally acceptable for near-surface disposal is waste for which form and disposal methods must be different, and in general more stringent, than those specified for Class C waste. In the absence of specific requirements in this part, such waste must be disposed of in a geologic repository as defined in part 60 of this chapter unless proposals for disposal of such waste in a disposal site licensed pursuant to this part are approved by the Commission.

---

[1]From the LEXIS® Online Service. Reformatted for this publication. Reprinted with the permission of LEXIS-NEXIS, a division of Reed Elsevier Inc.

(3) Classification determined by long-lived radionuclides. If radioactive waste contains only radionuclides listed in Table 1, classification shall be determined as follows:

(i) If the concentration does not exceed 0.1 times the value in Table 1, the waste is Class A.

(ii) If the concentration exceeds 0.1 times the value in Table 1 but does not exceed the value in Table 1, the waste is Class C.

(iii) If the concentration exceeds the value in Table 1, the waste is not generally acceptable for near-surface disposal.

(iv) For wastes containing mixtures of radionuclides listed in Table 1, the total concentration shall be determined by the sum of fractions rule described in paragraph (a)(7) of this section.

Table 1

Radionuclide	Concentration, curies per cubic meter
C–14	8
C–14 in activated metal	80
Ni–59 in activated metal	220
Nb–94 in activated metal	0.2
Tc–99	3
I–129	0.08
Alpha emitting transuranic nuclides with half-life greater than 5 years	[1]100
Pu–241	[1]3,500
Cm–242	[1]20,000

[1]Units are nanocuries per gram.

(4) Classification determined by short-lived radionuclides. If radioactive waste does not contain any of the radionuclides listed in Table 1, classification shall be determined based on the concentrations shown in Table 2. However, as specified in paragraph (a)(6) of this section, if radioactive waste does not contain any nuclides listed in either Table 1 or 2, it is Class A.

(i) If the concentration does not exceed the value in Column 1, the waste is Class A.

(ii) If the concentration exceeds the value in Column 1, but does not exceed the value in Column 2, the waste is Class B.

(iii) If the concentration exceeds the value in Column 2, but does not exceed the value in Column 3, the waste is Class C.

(iv) If the concentration exceeds the value in Column 3, the waste is not generally acceptable for near-surface disposal.

(v) For wastes containing mixtures of the nuclides listed in Table 2, the total concentration shall be determined by the sum of fractions rule described in paragraph (a)(7) of this section.

Table 2

Radionuclide	Concentration, curies per cubic meter Col. 1	Col. 2	Col. 3
Total of all nuclides with less than 5 year half-life	700	([1])	([1])
H-3	40	([1])	([1])
Co-60	700	([1])	([1])
Ni-63	3.5	70	700
Ni-63 in activated metal	35	700	7000
Sr-90	0.04	150	7000
Cs–137	1	44	4600

[1]There are no limits established for these radionuclides in Class B or C wastes. Practical considerations such as the effects of external radiation and internal heat generation on transportation, handling, and disposal will limit the concentrations for these wastes. These wastes shall be Class B unless the concentrations of other nuclides in Table 2 determine the waste to the Class C independent of these nuclides.

(5) Classification determined by both long- and short-lived radionuclides. If radioactive waste contains a mixture of radionuclides, some of which are listed in Table 1, and some of which are listed in Table 2, classification shall be determined as follows:

(i) If the concentration of a nuclide listed in Table 1 does not exceed 0.1 times the value listed in Table 1, the class shall be that determined by the concentration of nuclides listed in Table 2.

(ii) If the concentration of a nuclide listed in Table 1 exceeds 0.1 times the value listed in Table 1 but does not exceed the value in Table 1, the waste shall be Class C, provided the concentration of nuclides listed in Table 2 does not exceed the value shown in Column 3 of Table 2.

(6) Classification of wastes with radionuclides other than those listed in Tables 1 and 2. If radioactive waste does not contain any nuclides listed in either Table 1 or 2, it is Class A.

(7) The sum of the fractions rule for mixtures of radionuclides. For determining classification for waste that contains a mixture of radionuclides, it is necessary to determine the sum of fractions by dividing each nuclide's concentration by the appropriate limit and adding the resulting values. The appropriate limits must all be taken from the same column of the same table. The sum of the fractions for the column must be less than 1.0 if the waste class is to be determined by that column. Example: A waste contains Sr–90 in a concentration of 50 Ci/m^3 and Cs–137 in a concentration of 22 Ci/m^3. Since the concentrations both exceed the values in Column 1, Table 2, they must be compared to Column 2 values. For Sr–90 fraction 50/150 = 0.33; for Cs–137 fraction, 22/44 = 0.5; the sum of the fractions = 0.83. Since the sum is less than 1.0, the waste is Class B. . . .

Appendix H

Federal and New York State Low-Level Radioactive Waste Management Acts

1. Low-Level Radioactive Waste Policy Act. 1980 (P.L. 96-573, December 23). United States Statutes at Large 94:3347-3349.

2. Low-Level Radioactive Waste Policy Amendments Act of 1985. 1986 (P.L. 99-240, Title I, January 15, 1986). United States Statutes at Large 99:1842-1859. (From the LEXIS® Online Service. Reformatted for this publication. Reprinted with the permission of LEXIS-NEXIS, a division of Reed Elsevier Inc.)

3. New York State Low-Level Radioactive Waste Management Act. Laws of the State of New York, Chapter 673 (July 26):2801-2817.

1. LOW-LEVEL RADIOACTIVE WASTE POLICY ACT OF 1980

For Legislative History of this and other Laws, see Table 1, Public Laws and Legislative History, at end of final volume

An Act to set forth a Federal policy for the disposal of low-level radioactive wastes, and for other purposes.

Be it enacted by the Senate and House of Representatives of the United States of America in Congress assembled,

SHORT TITLE

SECTION 1. This Act may be cited as the "Low-Level Radioactive Waste Policy Act".

DEFINITIONS

SEC. 2. As used in this Act—

(1) The term "disposal" means the isolation of low-level radioactive waste pursuant to requirements established by the Nuclear Regulatory Commission under applicable laws.

(2) The term "low-level radioactive waste" means radioactive waste not classified as high-level radioactive waste, transuranic waste, spent nuclear fuel, or byproduct material as defined in section 11 e. (2) of the Atomic Energy Act of 1954.

(3) The term "State" means any State of the United States, the District of Columbia, and, subject to the provisions of Public Law 96–205, the Commonwealth of Puerto Rico, the Virgin Islands, Guam, the Northern Mariana Islands, the Trust Territory of the Pacific Islands, and any other territory or possession of the United States.

(4) For purposes of this Act the term "atomic energy defense activities of the Secretary" includes those activities and facilities of the Department of Energy carrying out the function of—

(i) Naval reactors development and propulsion,

(ii) weapons activities, verification and control technology,

(iii) defense materials production,

(iv) inertial confinement fusion,

(v) defense waste management, and

(vi) defense nuclear materials security and safeguards (all as included in the Department of Energy appropriations account in any fiscal year for atomic energy defense activities).

GENERAL PROVISIONS

SEC. 3. (a) Compacts established under this Act or actions taken under such compacts shall not be applicable to the transportation, management, or disposal of low-level radioactive waste from atomic energy defense activities of the Secretary or Federal research and development activities.

(b) Any facility established or operated exclusively for the disposal of low-level radioactive waste produced by atomic energy defense activities of the Secretary or Federal research and development activities shall not be subject to compacts established under this Act or actions taken under such compacts.

LOW-LEVEL RADIOACTIVE WASTE DISPOSAL

SEC. 4. (a)(1) It is the policy of the Federal Government that—

(A) each State is responsible for providing for the availability of capacity either within or outside the State for the disposal of low-level radioactive waste generated within its borders except for waste generated as a result of defense activities of the Secretary or Federal research and development activities; and

(B) low-level radioactive waste can be most safely and efficiently managed on a regional basis.

(2)(A) To carry out the policy set forth in paragraph (1), the States may enter into such compacts as may be necessary to provide for the establishment and operation of regional disposal facilities for low-level radioactive waste.

(B) A compact entered into under subparagraph (A) shall not take effect until the Congress has by law consented to the compact. Each such compact shall provide that every 5 years after the compact has taken effect the Congress may by law withdraw its consent. After January 1, 1986, any such compact may restrict the use of the regional disposal facilities under the compact to the disposal of low-level radioactive waste generated within the region.

(b)(1) In order to assist the States in carrying out the policy set forth in subsection (a)(1), the Secretary shall prepare and submit to Congress and to each of the States within 120 days after the date of the enactment of this Act a report which—

(A) defines the disposal capacity needed for present and future low-level radioactive waste on a regional basis;

(B) defines the status of all commercial low-level radioactive waste disposal sites and includes an evaluation of the license status of each such site, the state of operation of each site, including operating history, an analysis of the adequacy of disposal technology employed at each site to contain low-level radioactive wastes for their hazardous lifetimes, and such recommendations as the Secretary considers appropriate to assure protection of the public health and safety from wastes transported to such sites;

(C) evaluates the transportation requirements on a regional basis and in comparison with performance of present transportation practices for the shipment of low-level radioactive wastes, including an inventory of types and quantities of low-level wastes, and evaluation of shipment requirements for each type of waste and an evaluation of the ability of generators, shippers, and carriers to meet such requirements; and

(D) evaluates the capability of the low-level radioactive waste disposal facilities owned and operated by the Department of Energy to provide interim storage for commercially generated low-level waste and estimates the costs associated with such interim storage.

(2) In carrying out this subsection, the Secretary shall consult with the Governors of the States, the Nuclear Regulatory Commission, the Environmental Protection Agency, the United States Geological Survey, and the Secretary of Transportation, and such other agencies and departments as he finds appropriate.

Approved December 22, 1980.

LEGISLATIVE HISTORY:

SENATE REPORT No. 96-548 (Comm. on Energy and Natural Resources).
CONGRESSIONAL RECORD, Vol. 126 (1980):
July 28-30, considered and passed Senate.
Dec. 3, H.R. 8378 considered and passed House; passage vacated and S. 2189, amended, passed in lieu.
Dec. 13, Senate agreed to the House amendment with amendments; House agreed to Senate amendments.

2. LOW-LEVEL RADIOACTIVE WASTE POLICY AMENDMENTS ACT OF 1985

An Act

To amend the Low-Level Radioactive Waste Policy Act to improve procedures for the implementation of compacts providing for the establishment and operation of regional disposal facilities for low-level radioactive waste; to grant the consent of the Congress to certain interstate compacts on low-level radioactive waste; and for other purposes.

Be it enacted by the Senate and House of Representatives of the United States of America in Congress assembled,

TITLE I—LOW-LEVEL RADIOACTIVE WASTE POLICY AMENDMENTS ACT OF 1985

SEC. 101. SHORT TITLE.

This Title may be cited as the "Low-Level Radioactive Waste Policy Amendments Act of 1985".

SEC. 102. AMENDMENT TO THE LOW-LEVEL RADIOACTIVE WASTE POLICY ACT.

The Low-Level Radioactive Waste Policy Act (42 U.S.C. 2021b et seq.) is amended by striking out sections 1, 2, 3, and 4 and inserting in lieu thereof the following:

"SECTION 1. SHORT TITLE.

"This Act may be cited as the 'Low-Level Radioactive Waste Policy Act'.

"SEC. 2. DEFINITIONS.

"For purposes of this Act:

"(1) AGREEMENT STATE—The term 'agreement State' means a State that—

"(A) has entered into an agreement with the Nuclear Regulatory Commission under section 274 of the Atomic Energy Act of 1954 (42 U.S.C. 2021); and

"(B) has authority to regulate the disposal of low-level radioactive waste under such agreement.

"(2) ALLOCATION.—The term 'allocation' means the assignment of a specific amount of low-level radioactive waste disposal capacity to a commercial nuclear power reactor for which access is required to be provided by sited States subject to the conditions specified under this Act.

"(3) COMMERCIAL NUCLEAR POWER REACTOR.—The term 'commercial nuclear power reactor' means any unit of a civilian light-water moderated utilization facility required to be licensed under section 103 or 104b. of the Atomic Energy Act of 1954 (42 U.S.C. 2133 or 2134(b)).

"(4) COMPACT.—The term 'compact' means a compact entered into by two or more States pursuant to this Act.

"(5) COMPACT COMMISSION.—The term 'compact commission' means the regional commission, committee, or board established in a compact to administer such compact.

"(6) COMPACT REGION.—The term 'compact region' means the area consisting of all States that are members of a compact.

"(7) DISPOSAL.—The term 'disposal' means the permanent isolation of low-level radioactive waste pursuant to the requirements established by the Nuclear Regulatory Commission under applicable laws, or by an agreement State if such isolation occurs in such agreement State.

"(8) GENERATE.—The term 'generate', when used in relation to low-level radioactive waste, means to produce low-level radioactive waste.

"(9) LOW-LEVEL RADIOACTIVE WASTE.—The term 'low-level radioactive waste' means radioactive material that—

"(A) is not high-level radioactive waste, spent nuclear fuel, or byproduct material (as defined in section 11e.(2) of the Atomic Energy Act of 1954 (42 U.S.C. 2014(e)(2))); and

"(B) the Nuclear Regulatory Commission, consistent with existing law and in accordance with paragraph (A), classifies as low-level radioactive waste.

"(10) NON-SITED COMPACT REGION.—The term 'non-sited compact region' means any compact region that is not a sited compact region.

"(11) REGIONAL DISPOSAL FACILITY.—The term 'regional disposal facility' means a non-Federal low-level radioactive waste disposal facility in operation on January 1, 1985, or subsequently established and operated under a compact.

"(12) SECRETARY.—The term 'Secretary' means the Secretary of Energy.

"(13) SITED COMPACT REGION.—The term 'sited compact region' means a compact region in which there is located one of the regional disposal facilities at Barnwell, in the State of South Carolina; Richland, in the State of Washington; or Beatty, in the State of Nevada.

"(14) STATE.—The term 'State' means any State of the United States, the District of Columbia, and the Commonwealth of Puerto Rico.

"SEC. 3. RESPONSIBILITIES FOR DISPOSAL OF LOW-LEVEL RADIOACTIVE WASTE.

"SECTION 3(a)(1) STATE RESPONSIBILITIES.—Each State shall be responsible for providing, either by itself or in cooperation with other States, for the disposal of—

"(A) low-level radioactive waste generated within the State (other than by the Federal Government) that consists of or contains class A, B, or C radioactive waste as defined by section 61.55 of title 10, Code of Federal Regulations, as in effect on January 26, 1983;

"(B) low-level radioactive waste described in subparagraph (A) that is generated by the Federal Government except such waste that is—

"(i) owned or generated by the Department of Energy;

"(ii) owned or generated by the United States Navy as a result of the decommissioning of vessels of the United States Navy; or

"(iii) owned or generated as a result of any research, development, testing, or production of any atomic weapon; and

"(C) low-level radioactive waste described in subparagraphs (A) and (B) that is generated outside of the State and accepted for disposal in accordance with sections 5 or 6.

"(2) No regional disposal facility may be required to accept for disposal any material—

"(A) that is not low-level radioactive waste as defined by section 61.55 of title 10, Code of Federal Regulations, as in effect on January 26, 1983, or

"(B) identified under the Formerly Utilized Sites Remedial Action Program.

Nothing in this paragraph shall be deemed to prohibit a State, subject to the provisions of its compact, or a compact region from accepting for disposal any material identified in subparagraph (A) or (B).

"(b)(1) The Federal Government shall be responsible for the disposal of—

"(A) low-level radioactive waste owned or generated by the Department of Energy;

"(B) low-level radioactive waste owned or generated by the United States Navy as a result of the decommissioning of vessels of the United States Navy;

"(C) low-level radioactive waste owned or generated by the Federal Government as a result of any research, development, testing, or production of any atomic weapon; and

"(D) any other low-level radioactive waste with concentrations of radionuclides that exceed the limits established by the Commission for class C radioactive waste, as defined by section 61.55 of title 10, Code of Federal Regulations, as in effect on January 26, 1983.

"(2) All radioactive waste designated a Federal responsibility pursuant to subparagraph (b)(1)(D) that results from activities licensed by the Nuclear Regulatory Commission under the Atomic Energy Act of 1954, as amended, shall be disposed of in a facility licensed by the Nuclear Regulatory Commission that the Commission determines is adequate to protect the public health and safety.

"(3) Not later than 12 months after the date of enactment of this Act, the Secretary shall submit to the Congress a comprehensive report setting forth the recommendations of the Secretary for ensuring the safe disposal of all radioactive waste designated a Federal responsibility pursuant to subparagraph (b)(1)(D). Such report shall include—

"(A) an identification of the radioactive waste involved, including the source of such waste, and the volume, concentration, and other relevant characteristics of such waste;

"(B) an identification of the Federal and non-Federal options for disposal of such radioactive waste;

"(C) a description of the actions proposed to ensure the safe disposal of such radioactive waste;

"(D) a description of the projected costs of undertaking such actions;

"(E) an identification of the options for ensuring that the beneficiaries of the activities resulting in the generation of such radioactive wastes bear all reasonable costs of disposing of such wastes; and

"(F) an identification of any statutory authority required for disposal of such waste.

"(4) The Secretary may not dispose of any radioactive waste designated a Federal responsibility pursuant to paragraph (b)(1)(D) that becomes a Federal responsibility for the first time pursuant to such paragraph until ninety days after the report prepared pursuant to paragraph (3) has been submitted to the Congress.

"SEC. 4. REGIONAL COMPACTS FOR DISPOSAL OF LOW-LEVEL RADIOACTIVE WASTE.

"(a) IN GENERAL.—

"(1) FEDERAL POLICY.—It is the policy of the Federal Government that the responsibilities of the States under section 3 for the disposal of low-level radioactive waste can be most safely and effectively managed on a regional basis.

"(2) INTERSTATE COMPACTS.—To carry out the policy set forth in paragraph (1), the States may enter into such compacts as may be necessary to provide for the establishment and operation of regional disposal facilities for low-level radioactive waste.

"(b) APPLICABILITY TO FEDERAL ACTIVITIES.—

"(1) IN GENERAL.—

"(A) ACTIVITIES OF THE SECRETARY.—Except as provided in subparagraph (B), no compact or action taken under a compact shall be applicable to the transportation, management, or disposal of any low-level radioactive waste designated in section 3(a)(1)(B) (i)-(iii).

"(B) FEDERAL LOW-LEVEL RADIOACTIVE WASTE DISPOSED OF AT NON-FEDERAL FACILITIES.—Low-level radioactive waste owned or generated by the Federal Government that is disposed of at a regional disposal facility or non-Federal disposal facility within a State that is not a member of a compact shall be subject to the same conditions, regulations, requirements, fees, taxes, and surcharges imposed by the compact commission, and by the State in which such facility is located, in the same manner and to the same extent as any low-level radioactive waste not generated by the Federal Government.

"(2) FEDERAL LOW-LEVEL RADIOACTIVE WASTE DISPOSAL FACILITIES.—Any low-level radioactive waste disposal facility established or operated exclusively for the disposal of low-level radioactive waste owned or generated by the Federal Government shall not be subject to any compact or any action taken under a compact.

"(3) EFFECT OF COMPACTS ON FEDERAL LAW.—Nothing contained in this Act or any compact may be construed to confer any new authority on any compact commission or State—

"(A) to regulate the packaging, generation, treatment, storage, disposal, or transportation of low-level radioactive waste in a manner incompatible with the regulations of the Nuclear Regulatory Commission or inconsistent with the regulations of the Department of Transportation;

"(B) to regulate health, safety, or environmental hazards from source material, byproduct material, or special nuclear material;

"(C) to inspect the facilities of licensees of the Nuclear Regulatory Commission;

"(D) to inspect security areas or operations at the site of the generation of any low-level radioactive waste by the Federal Government, or to inspect classified information related to such areas or operations; or

"(E) to require indemnification pursuant to the provisions of chapter 171 of title 28, United States Code (commonly referred to as the Federal Tort Claims Act), or section 170 of the Atomic Energy Act of 1954 (42 U.S.C. 2210) (commonly referred to as the Price-Anderson Act), whichever is applicable.

"(4) FEDERAL AUTHORITY.—Except as expressly provided in this Act, nothing contained in this Act or any compact may be construed to limit the applicability of any Federal law or to diminish or otherwise impair the jurisdiction of any Federal agency, or to alter, amend, or otherwise affect any Federal law governing the judicial review of any action taken pursuant to any compact.

"(5) STATE AUTHORITY PRESERVED.—Except as expressly provided in this Act, nothing contained in this Act expands, diminishes, or otherwise affects State law.

"(c) Restricted Use of Regional Disposal Facilities.—Any authority in a compact to restrict the use of the regional disposal facilities under the compact to the disposal of low-level radioactive waste generated within the compact region shall not take effect before each of the following occurs:

"(1) January 1, 1986; and

"(2) the Congress by law consents to the compact.

"(d) Congressional Review.—Each compact shall provide that every 5 years after the compact has taken effect the Congress may by law withdraw its consent.

"SEC. 5. LIMITED AVAILABILITY OF CERTAIN REGIONAL DISPOSAL FACILITIES DURING TRANSITION AND LICENSING PERIODS.

"(a) AVAILABILITY OF DISPOSAL CAPACITY.—

"(1) PRESSURIZED-WATER AND BOILING WATER REACTORS.—During the seven-year period beginning January 1, 1986 and ending December 31, 1992, subject to the provisions of subsections (b) through (g), each State in which there is located a regional disposal facility referred to in paragraphs (1) through (3) of subsection (b) shall make disposal capacity available for low-level radioactive waste generated by pressurized water and boiling water commercial nuclear power reactors in accordance with the allocations established in subsection (c).

"(2) OTHER SOURCES OF LOW-LEVEL RADIOACTIVE WASTE.—During the seven-year period beginning January 1, 1986 and ending December 31, 1992, subject to the provisions of subsections (b) through (g), each State in which there is located a regional disposal facility referred to in paragraphs (1) through (3) of subsection (b) shall make disposal capacity available for low-level radioactive waste generated by any source not referred to in paragraph (1).

"(3) ALLOCATION OF DISPOSAL CAPACITY.—

"(A) During the seven-year period beginning January 1, 1986 and ending December 31, 1992, low-level radioactive waste generated within a sited compact region shall be accorded priority under this section in the allocation of available disposal capacity at a regional disposal facility referred to in paragraphs (1) through (3) of subsection (b) and located in the sited compact region in which such waste is generated.

"(B) Any State in which a regional disposal facility referred to in paragraphs (1) through (3) of subsection (b) is located may, subject to the provisions of its compact, prohibit the disposal at such facility of low-level radioactive waste generated outside of the compact region if the disposal of such waste in any given calendar year, together with all other low-level radioactive waste disposed of at such facility within that same calendar year, would result in that facility disposing of a total annual volume of low-level radioactive waste in excess of 100 per centum of the average annual volume for such facility designated in subsection (b): Provided, however, That in the event that all three States in which regional disposal facilities referred to in paragraphs (1) through (3) of subsection (b) act to prohibit the disposal of low-level radioactive waste pursuant to this subparagraph, each such State shall, in accordance with any applicable procedures of its compact, permit, as necessary, the disposal of additional quantities of such waste in increments of 10 per centum of the average annual volume for each such facility designated in subsection (b).

"(C) Nothing in this paragraph shall require any disposal facility or State referred to in paragraphs (1) through (3) of subsection (b) to accept for disposal low-level radioactive waste in excess of the total amounts designated in subsection (b).

"(4) CESSATION OF OPERATION OF LOW-LEVEL RADIOACTIVE WASTE DISPOSAL FACILITY.— No provision of this section shall be construed to obligate any State referred to in paragraphs (1) through (3) of subsection (b) to accept low-level radioactive waste from any source in the event that the regional disposal facility located in such State ceases operations.

"(b) LIMITATIONS.—The availability of disposal capacity for low-level radioactive waste from any source shall be subject to the following limitations:

"(1) BARNWELL, SOUTH CAROLINA.—The State of South Carolina, in accordance with the provisions of its compact, may limit the volume of low-level radioactive waste accepted for disposal at the regional disposal facility located at Barnwell, South Carolina to a total of 8,400,000 cubic feet of low-level radioactive waste during the 7-year period beginning January 1, 1986, and ending December 31, 1992 (as based on an average annual volume of 1,200,000 cubic feet of low-level radioactive waste).

"(2) RICHLAND, WASHINGTON.—The State of Washington, in accordance with the provisions of its compact, may limit the volume of low-level radioactive waste accepted for disposal at the regional disposal facility located at Richland, Washington to a total of 9,800,000 cubic feet of low-level radioactive waste during the 7-year period beginning January 1, 1986, and ending December 31, 1992 (as based on an average annual volume of 1,400,000 cubic feet of low-level radioactive waste).

"(3) BEATTY, NEVADA.—The State of Nevada, in accordance with the provisions of its compact, may limit the volume of low-level radioactive waste accepted for disposal at the regional disposal facility located at Beatty, Nevada to a total of 1,400,000 cubic feet of low-level radioactive waste during the 7-year period beginning January 1, 1986, and ending December 31, 1992 (as based on an average annual volume of 200,000 cubic feet of low-level radioactive waste).

"(c) COMMERCIAL NUCLEAR POWER REACTOR ALLOCATIONS.—

"(1) AMOUNT.—Subject to the provisions of subsections (a) through (g) each commercial nuclear power reactor shall upon request receive an allocation of low-level radioactive waste disposal capacity (in cubic feet) at the facilities referred to in subsection (b) during the 4-year transition period beginning January 1, 1986, and ending December 31, 1989, and during the 3-year licensing period beginning January 1, 1990, and ending December 31, 1992, in an amount calculated by multiplying the appropriate number from the following table by the number of months remaining in the applicable period as determined under paragraph (2).

"Reactor Type	4-year Transition Period		3-year Transition Period	
	In Sited Region	All Other Locations	In Sited Region	All Other Locations
PWR................	1027	871	934	685
BWR...............	2300	1951	2091	1533

"(2) METHOD OF CALCULATION.—For purposes of calculating the aggregate amount of disposal capacity available to a commercial nuclear power reactor under this subsection, the number of months shall be computed beginning with the first month of the applicable period, or the sixteenth month after receipt of a full power operating license, whichever occurs later.

"(3) UNUSED ALLOCATIONS.—Any unused allocation under paragraph (1) received by a reactor during the transition period or the licensing period may be used at any time after such reactor receives its full power license or after the beginning of the pertinent period, whichever is later, but not in any event after December 31, 1992, or after commencement of operation of a regional disposal facility in the compact region or State in which such reactor is located, whichever occurs first.

"(4) TRANSFERABILITY.—Any commercial nuclear power reactor in a State or compact region that is in compliance with the requirements of subsection (e) may assign any disposal capacity allocated to it under this subsection to any other person in each State or compact region. Such assignment may be for valuable consideration and shall be in writing, copies of which shall be filed at the affected compact commissions and States, along with the assignor's unconditional written waiver of the disposal capacity being assigned.

"(5) UNUSUAL VOLUMES.—

"(A) The Secretary may, upon petition by the owner or operator of any commercial nuclear power reactor, allocate to such reactor disposal capacity in excess of the amount calculated under paragraph (1) if the Secretary finds and states in writing his reasons for so finding that making additional capacity available for such reactor through this paragraph is required to permit unusual or unexpected operating, maintenance, repair or safety activities.

"(B) The Secretary may not make allocations pursuant to subparagraph (A) that would result in the acceptance for disposal of more than 800,000 cubic feet of low-level radioactive waste or would result in the total of the allocations made pursuant to this subsection exceeding 11,900,000 cubic feet over the entire seven-year interim access period.

"(6) LIMITATION.—During the seven-year interim access period referred to in subsection (a), the disposal facilities referred to in subsection (b) shall not be required to accept more than 11,900,000 cubic feet of low-level radioactive waste generated by commercial nuclear power reactors.

"(d)(1) SURCHARGES.—The disposal of any low-level radioactive waste under this section (other than low-level radioactive waste generated in a sited compact region) may be charged a surcharge by the State in which the applicable regional disposal facility is located, in addition to the fees and surcharges generally applicable for disposal of low-level radioactive waste in the regional disposal facility involved. Except as provided in subsection (e)(2), such surcharges shall not exceed—

"(A) in 1986 and 1987, $10 per cubic foot of low-level radioactive waste;

"(B) in 1988 and 1989, $20 per cubic foot of low-level radioactive waste; and

"(C) in 1990, 1991, and 1992, $40 per cubic foot of low-level radioactive waste.

"(2) MILESTONE INCENTIVES.—

"(A) ESCROW ACCOUNT.—Twenty-five per centum of all surcharge fees received by a State pursuant to paragraph (1) during the seven-year period referred to in subsection (a) shall be transferred on a monthly basis to an escrow account held by the Secretary. The Secretary shall deposit all funds received in a special escrow account. The funds so deposited shall not be the property of the United States. The Secretary shall act as trustee for such funds and shall invest them in interest-bearing United States Government Securities with the highest available yield. Such funds shall be held by the Secretary until—

"(i) paid or repaid in accordance with subparagraph (B) or (C); or "(ii) paid to the State collecting such fees in accordance with subparagraph (F).

"(B) PAYMENTS.—

"(i) July 1, 1986.—The twenty-five per centum of any amount collected by a State under paragraph (1) for low-level radioactive waste disposed of under this section during the period beginning on the date of enactment of the Low-Level Radioactive Waste Policy Amendments Act of 1985 and ending June 30, 1986, and transferred to the Secretary under subparagraph (A), shall be paid by the Secretary in accordance with subparagraph (D) if the milestone described in subsection (e)(1)(A) is met by the State in which such waste originated.

"(ii) January 1, 1988.—The twenty-five per centum of any amount collected by a State under paragraph (1) for low-level radioactive waste disposed of under this section during the period beginning July 1, 1986 and ending December 31, 1987, and transferred to the Secretary under subparagraph (A), shall be paid by the Secretary in accordance with subparagraph (D) if the milestone described in subsection (e)(1)(B) is met by the State in which such waste originated (or its compact region, where applicable).

"(iii) January 1, 1990.—The twenty-five per centum of any amount collected by a State under paragraph (1) for low-level radioactive waste disposed of under this section during the period beginning January 1, 1988 and ending December 31, 1989, and transferred to the Secretary under subparagraph (A), shall be paid by the Secretary in accordance with subparagraph (D) if the milestone described in subsection (e)(1)(C) is met by the State in which such waste originated (or its compact region, where applicable).

"(iv) The twenty-five per centum of any amount collected by a State under paragraph (1) for low-level radioactive waste disposed of under this section during the period beginning January 1, 1990 and ending December 31, 1992, and transferred to the Secretary under subparagraph (A), shall be paid by the Secretary in accordance with subparagraph (D) if, by January 1, 1993, the State in which such waste originated (or its compact region, where applicable) is able to provide for the disposal of all low-level radioactive waste generated within such State or compact region.

"(C) FAILURE TO MEET JANUARY 1, 1993 DEADLINE.—If, by January 1, 1993, a State (or, where applicable, a compact region) in which low-level radioactive waste is generated is unable to provide for the disposal of all such waste generated within such State or compact region—

"(i) each State in which such waste is generated, upon the request of the generator or owner of the waste, shall take title to the waste, shall be obligated to take possession of the waste, and shall be liable for all damages directly or indirectly incurred by such generator or owner as a consequence of the failure of the State to take possession of the waste as soon after January 1, 1993 as the generator or owner notifies the State that the waste is available for shipment; or

"(ii) if such State elects not to take title to, take possession of, and assume liability for such waste, pursuant to clause (I), twenty-five per centum of any amount collected by a State under paragraph (1) for low-level radioactive waste disposed of under this section during the period beginning January 1, 1990 and ending December 31, 1992 shall be repaid, with interest, to each generator from whom such surcharge was collected. Repayments made pursuant to this clause shall be made on a monthly basis, with the first such repayment beginning on February 1, 1993, in an amount equal to one thirty-sixth of the total amount required to be repaid pursuant to this clause, and shall continue until the State (or, where applicable, compact region) in which such low-level radioactive waste is generated is able to provide for the disposal of all such waste generated within such State or compact region or until January 1, 1996, whichever is earlier.

If a State in which low-level radioactive waste is generated elects to take title to, take possession of, and assume liability for such waste pursuant to clause (i), such State shall be paid such amounts as are designated in subparagraph (B)(iv). If a State (or, where applicable, a compact region) in which low-level radioactive waste is generated provides for the disposal of such waste at any time after January 1, 1993 and prior to January 1, 1996, such State (or, where applicable, compact region) shall be paid in accordance with subparagraph (D) a lump sum amount equal to twenty-five per centum of any amount collected by a State under paragraph (1): Provided, however, That such payment shall be adjusted to reflect the remaining number of months between January 1, 1993 and January 1, 1996 for which such State (or, where applicable, compact region) provides for the disposal of such waste. If a State (or, where applicable, a compact region) in which low-level radioactive waste is generated is unable to provide for the disposal of all such waste generated within such State or compact region by January 1, 1996, each State in which such waste is generated, upon the request of the generator or owner of the waste, shall take title to the waste, be obligated to take possession of the waste, and shall be liable for all damages directly or indirectly incurred by such generator or owner as a consequence of the failure of the State to take possession of the waste as soon after January 1, 1996, as the generator or owner notifies the State that the waste is available for shipment.

"(D) RECIPIENTS OF PAYMENTS.—The payments described in subparagraphs (B) and (C) shall be paid within thirty days after the applicable date—

"(i) if the State in which such waste originated is not a member of a compact region, to such State;

"(ii) if the State in which such waste originated is a member of the compact region, to the compact commission serving such State.

"(E) USES OF PAYMENTS.—

"(i) LIMITATIONS.—Any amount paid under subparagraphs (B) or (C) may only be used to—

"(I) establish low-level radioactive waste disposal facilities;

"(II) mitigate the impact of low-level radioactive waste disposal facilities on the host State;

"(III) regulate low-level radioactive waste disposal facilities; or

"(IV) ensure the decommissioning, closure, and care during the period of institutional control of low-level radioactive waste disposal facilities.

"(ii) REPORTS.—

"(I) RECIPIENT.—Any State or compact commission receiving a payment under subparagraphs (B) or (C) shall, on December 31 of each year in which any such funds are expended, submit a report to the Department of Energy itemizing any such expenditures.

"(II) DEPARTMENT OF ENERGY.—Not later than six months after receiving the reports under subclause (I), the Secretary shall submit to the Congress a summary of all such reports that shall include an assessment of the compliance of each such State or compact commission with the requirements of clause (i).

"(F) PAYMENT TO STATES.—Any amount collected by a State under paragraph (1) that is placed in escrow under subparagraph (A) and not paid to a State or compact commission under subparagraphs (B) and (C) or not repaid to a generator under subparagraph (C) shall be paid from such escrow account to such State collecting such payment under paragraph (1). Such payment shall be made not later than 30 days after a determination of ineligibility for a refund is made.

"(G) PENALTY SURCHARGES.—No rebate shall be made under this subsection of any surcharge or penalty surcharge paid during a period of noncompliance with subsection (e)(1).

"(e) REQUIREMENTS FOR ACCESS TO REGIONAL DISPOSAL FACILITIES.—

"(1) REQUIREMENTS FOR NON-SITED COMPACT REGIONS AND NON-MEMBER STATES.—Each non-sited compact region, or State that is not a member of a compact region that does not have an operating disposal facility, shall comply with the following requirements:

"(A) By July 1, 1986, each such non-member State shall ratify compact legislation or, by the enactment of legislation or the certification of the Governor, indicate its intent to develop a site for the location of a low-level radioactive waste disposal facility within such State.

"(B) By January 1, 1988.—

"(i) each non-sited compact region shall identify the State in which its low-level radioactive waste disposal facility is to be located, or shall have selected the developer for such facility and the site to be developed, and each compact region or the State in which its low-level radioactive waste disposal facility is to be located shall develop a siting plan for such facility providing detailed procedures and a schedule for establishing a facility location and preparing a facility license application and shall delegate authority to implement such plan;

"(ii) each non-member State shall develop a siting plan providing detailed procedures and a schedule for establishing a facility location and preparing a facility license application for a low-level radioactive waste disposal facility and shall delegate authority to implement such plan; and

"(iii) The siting plan required pursuant to this paragraph shall include a description of the optimum way to attain operation of the low-level radioactive waste disposal facility involved, within the time period specified in this Act. Such plan shall include a description of the objectives and a sequence of deadlines for all entities required to take action to implement such plan, including, to the extent practicable, an identification of the activities in which a delay in the start, or completion, of such activities will cause a delay in beginning facility operation. Such plan shall also identify, to the extent practicable, the process for (1) screening for broad siting areas; (2) identifying and evaluating specific candidate sites; and (3) characterizing the preferred site(s), completing all necessary environmental assessments, and preparing a license application for submission to the Nuclear Regulatory Commission or an Agreement State.

"(C) By January 1, 1990.—

"(i) a complete application (as determined by the Nuclear Regulatory Commission or the appropriate agency of an agreement State) shall be filed for a license to operate a low-level radioactive waste disposal facility within each non-sited compact region or within each non-member State; or

"(ii) the Governor (or, for any State without a Governor, the chief executive officer) of any State that is not a member of a compact region in compliance with clause (i), or has not complied with such clause by its own actions, shall provide a written certification to the Nuclear Regulatory Commission, that such State will be capable of providing for, and will provide for, the storage, disposal, or management of any low-level radioactive waste generated within such State and requiring disposal after December 31, 1992, and include a description of the actions that will be taken to ensure that such capacity exists.

"(D) By January 1, 1992, a complete application (as determined by the Nuclear Regulatory Commission or the appropriate agency of an agreement State) shall be filed for a license to operate a low-level radioactive waste disposal facility within each non-sited compact region or within each non-member State.

"(E) The Nuclear Regulatory Commission shall transmit any certification received under subparagraph (C) to the Congress and publish any such certification in the Federal Register.

"(F) Any State may, subject to all applicable provisions, if any, of any applicable compact, enter into an agreement with the compact commission of a region in which a regional disposal facility is located to provide for the disposal of all low-level radioactive waste generated within such State, and, by virtue of such agreement, may, with the approval of the State in which the regional disposal facility is located, be deemed to be in compliance with subparagraphs (A), (B), (C), and (D).

"(2) PENALTIES FOR FAILURE TO COMPLY.—

"(A) BY JULY 1, 1986.—If any State fails to comply with subparagraph (1)(A)—

"(i) any generator of low-level radioactive waste within such region or non-member State shall, for the period beginning July 1, 1986, and ending December 31, 1986, be charged 2 times the surcharge otherwise applicable under subsection (d); and

"(ii) on or after January 1, 1987, any low-level radioactive waste generated within such region or non-member State may be denied access to the regional disposal facilities referred to in paragraphs (1) through (3) of subsection (b).

"(B) BY JANUARY 1, 1988.—If any non-sited compact region or non-member State fails to comply with paragraph (1)(B)—

"(i) any generator of low-level radioactive waste within such region or non-member State shall—

"(I) for the period beginning January 1, 1988, and ending June 30, 1988, be charged 2 times the surcharge otherwise applicable under subsection (d); and

"(II) for the period beginning July 1, 1988, and ending December 31, 1988, be charged 4 times the surcharge otherwise applicable under subsection (d); and

"(ii) on or after January 1, 1989, any low-level radioactive waste generated within such region or non-member State may be denied access to the regional disposal facilities referred to in paragraphs (1) through (3) of subsection (b).

"(C) BY JANUARY 1, 1990.—If any non-sited compact region or non-member State fails to comply with paragraph (1)(C), any low-level radioactive waste generated within such region or non-member State may be denied access to the regional disposal facilities referred to in paragraphs (1) through (3) of subsection (b).

"(D) BY JANUARY 1, 1992.—If any non-sited compact region or non-member State fails to comply with paragraph (1)(D), any generator of low-level radioactive waste within such region or non-member State shall, for the period beginning January 1, 1992 and ending upon the filing of the application described in paragraph (1)(D), be charged 3 times the surcharge otherwise applicable under subsection (d).

"(3) DENIAL OF ACCESS.—No denial or suspension of access to a regional disposal facility under paragraph (2) may be based on the source, class, or type of low-level radioactive waste.

"(4) RESTORATION OF SUSPENDED ACCESS; PENALTIES FOR FAILURE TO COMPLY.—Any access to a regional disposal facility that is suspended under paragraph (2) shall be restored after the non-sited compact region or non-member State involved complies with such requirement. Any payment of surcharge penalties pursuant to paragraph (2) for failure to comply with the requirements of subsection (e) shall be terminated after the non-sited compact region or non-member State involved complies with such requirements.

"(f)(1) ADMINISTRATION.—Each State and compact commission in which a regional disposal facility referred to in paragraphs (1) through (3) of subsection (b) is located shall have authority—

"(A) to monitor compliance with the limitations, allocations, and requirements established in this section; and

"(B) to deny access to any non-Federal low-level radioactive waste disposal facilities within its borders to any low-level radioactive waste that—

"(i) is in excess of the limitations or allocations established in this section; or

"(ii) is not required to be accepted due to the failure of a compact region or State to comply with the requirements of subsection (e)(1).

"(2) AVAILABILITY OF INFORMATION DURING INTERIM ACCESS PERIOD.—

"(A) The States of South Carolina, Washington, and Nevada may require information from disposal facility operators, generators, intermediate handlers, and the Department of Energy that is reasonably necessary to monitor the availability of disposal capacity, the use and assignment of allocations and the applicability of surcharges.

"(B) The States of South Carolina, Washington, and Nevada may, after written notice followed by a period of at least 30 days, deny access to disposal capacity to any generator or intermediate handler who fails to provide information under subparagraph (A).

"(C) PROPRIETARY INFORMATION.—

"(i) Trade secrets, proprietary and other confidential information shall be made available to a State under this subsection upon request only if such State—

"(I) consents in writing to restrict the dissemination of the information to those who are directly involved in monitoring under subparagraph (A) and who have a need to know;

"(II) accepts liability for wrongful disclosure; and

"(III) demonstrates that such information is essential to such monitoring.

"(ii) The United States shall not be liable for the wrongful disclosure by any individual or State of any information provided to such individual or State under this subsection.

"(iii) Whenever any individual or State has obtained possession of information under this subsection, the individual shall be subject to the same provisions of law with respect to the disclosure of such information as would apply to an officer or employee of the United States or of any department or agency thereof and the State shall be subject to the same provisions of law with respect to the disclosure of such information as would apply to the United States or any department or agency thereof. No State or State officer or employee who receives trade secrets, proprietary information, or other confidential information under this Act may be required to disclose such information under State law.

"(g) NONDISCRIMINATION.—Except as provided in subsections (b) through (e), low-level radioactive waste disposed of under this section shall be subject without discrimination to all applicable legal requirements of the compact region and State in which the disposal facility is located as if such low-level radioactive waste were generated within such compact region.

"SEC. 6. EMERGENCY ACCESS.

"(a) IN GENERAL.—The Nuclear Regulatory Commission may grant emergency access to any regional disposal facility or non-Federal disposal facility within a State that is not a member of a compact for specific low-level radioactive waste, if necessary to eliminate an immediate and serious threat to the public health and safety or the common defense and security. The procedure for granting emergency access shall be as provided in this section.

"(b) REQUEST FOR EMERGENCY ACCESS.—Any generator of low-level radioactive waste, or any Governor (or, for any State without a Governor, the chief executive officer of the State) on behalf of any generator or generators located in his or her State, may request that the Nuclear Regulatory Commission grant emergency access to a regional disposal facility or a non-Federal disposal facility within a State that is not a member of a compact for specific low-level radioactive waste. Any such request shall contain any information and certifications the Nuclear Regulatory Commission may require.

"(c) DETERMINATION OF NUCLEAR REGULATORY COMMISSION.—

"(1) REQUIRED DETERMINATION.—Not later than 45 days after receiving a request under subsection (b), the Nuclear Regulatory Commission shall determine whether—

"(A) emergency access is necessary because of an immediate and serious threat to the public health and safety or the common defense and security; and

"(B) the threat cannot be mitigated by any alternative consistent with the public health and safety, including storage of low-level radioactive waste at the site of generation or in a storage facility obtaining access to a disposal facility by voluntary agreement, purchasing disposal capacity available for assignment pursuant to section 5(c) or ceasing activities that generate low-level radioactive waste.

"(2) REQUIRED NOTIFICATION.—If the Nuclear Regulatory Commission makes the determinations required in paragraph (1) in the affirmative, it shall designate an appropriate non-Federal disposal facility or facilities, and notify the Governor (or chief executive officer) of the State in which such facility is located and the appropriate compact commission that emergency access is required. Such notification shall specifically describe the low-level radioactive waste as to source, physical and radiological characteristics, and the minimum volume and duration, not exceeding 180 days, necessary to alleviate the immediate threat to public health and safety or the common defense and security. The Nuclear Regulatory Commission shall also notify the Governor (or chief executive officer) of the State in which the low-level radioactive waste requiring emergency access was generated that emergency access has been granted and that, pursuant to subsection (e), no extension of emergency access may be granted absent diligent State action during the period of the initial grant.

"(d) TEMPORARY EMERGENCY ACCESS.—Upon determining that emergency access is necessary because of an immediate and serious threat to the public health and safety or the common defense and security, the Nuclear Regulatory Commission may at its discretion grant temporary emergency access, pending its determination whether the threat could be mitigated by any alternative consistent with the public health and safety. In granting access under this subsection, the Nuclear Regulatory Commission shall provide the same notification and information required under subsection (c). Absent a determination that no alternative consistent with the public health and safety would mitigate the threat, access granted under this subsection shall expire 45 days after the granting of temporary emergency access under this subsection.

"(e) EXTENSION OF EMERGENCY ACCESS.—The Nuclear Regulatory Commission may grant one extension of emergency access beyond the period provided in subsection (c), if it determines that emergency access continues to be necessary because of an immediate and serious threat to the public health and safety or the common defense and security that cannot be mitigated by any alternative consistent with the public health and safety, and that the generator of low-level radioactive waste granted emergency access and the State in which such low-level radioactive waste was generated have diligently though unsuccessfully acted during the period of the initial grant to eliminate the need for emergency access. Any extension granted under this subsection shall be for the minimum volume and duration the Nuclear Regulatory Commission finds necessary to eliminate the immediate threat to public health and safety or the common defense and security, and shall not in any event exceed 180 days.

"(f) RECIPROCAL ACCESS.—Any compact region or State not a member of a compact that provides emergency access to non-Federal disposal facilities within its borders shall be entitled to reciprocal access to any subsequently operating non-Federal disposal facility that serves the State or compact region in which low-level radioactive waste granted emergency access was generated. The compact commission or State having authority to approve importation of low-level radioactive waste to the disposal facility to which emergency access was granted shall designate for reciprocal access an equal volume of low-level radioactive waste having similar characteristics to that provided emergency access.

"(g) APPROVAL BY COMPACT COMMISSION.—Any grant of access under this section shall be submitted to the compact commission for the region in which the designated disposal facility is located for such approval as may be required under the terms of its compact. Any such compact commission shall act to approve emergency access not later than 15 days after receiving notification from the Nuclear Regulatory Commission, or reciprocal access not later than 15 days after receiving notification from the appropriate authority under subsect (f).

"(h) LIMITATIONS.—No State shall be required to provide emergency or reciprocal access to any regional disposal facility within its borders for low-level radioactive waste not meeting criteria established by the license or license agreement of such facility, or in excess of the approved capacity of such facility, or to delay the closing of any such facility pursuant to plans established before receiving a request for emergency or reciprocal access. No State shall, during any 12-month period, be required to provide emergency or reciprocal access to any regional disposal facility within its borders for more than 20 percent of the total volume of low-level radioactive waste accepted for disposal at such facility during the previous calendar year.

"(i) VOLUME REDUCTION AND SURCHARGES.—Any low-level radioactive waste delivered for disposal under this section shall be reduced in volume to the maximum extent practicable and shall be subject to surcharges established in this Act.

"(j) DEDUCTION FROM ALLOCATION.—Any volume of low-level radioactive waste granted emergency or reciprocal access under this section, if generated by any commercial nuclear power reactor, shall be deducted from the low-level radioactive waste volume allocable under section 5(c).

"(k) AGREEMENT STATES.—Any agreement under section 274 of the Atomic Energy Act of 1954 (42 U.S.C. 2021) shall not be applicable to the determinations of the Nuclear Regulatory Commission under this section.

"SEC. 7. RESPONSIBILITIES OF THE DEPARTMENT OF ENERGY.

"(a) FINANCIAL AND TECHNICAL ASSISTANCE.—The Secretary shall, to the extent provided in appropriations Act, provide to those compact regions, host States, and nonmember States determined by the Secretary to require assistance for purposes of carrying out this Act—

"(1) continuing technical assistance to assist them in fulfilling their responsibilities under this Act. Such technical assistance shall include, but not be limited to, technical guidelines for site selection, alternative technologies for low-level radioactive waste disposal, volume reduction options, management techniques to reduce low-level waste generation, transportation practices for shipment of low-level wastes, health and safety considerations in the storage, shipment and disposal of low-level radioactive wastes, and establishment of a computerized data-base to monitor the management of low-level radioactive wastes; and

"(2) through the end of fiscal year 1993, financial assistance to assist them in fulfilling their responsibilities under this Act.

"(b) REPORTS.—The Secretary shall prepare and submit to the Congress on an annual basis a report which (1) summarizes the progress of low-level waste disposal siting and licensing activities within each compact region, (2) reviews the available volume reduction technologies, their applications, effectiveness, and costs on a per unit volume basis, (3) reviews interim storage facility requirements, costs, and usage, (4) summarizes transportation requirements for such wastes on an inter- and intra-regional basis, (5) summarizes the data on the total amount of low-level waste shipped for disposal on a yearly basis, the proportion of such wastes subjected to volume reduction, the average volume reduction attained, and the proportion of wastes stored on an interim basis, and

(6) projects the interim storage and final disposal volume requirements anticipated for the following year, on a regional basis.

"SEC. 8. ALTERNATIVE DISPOSAL METHODS.

"(a) Not later than 12 months after the date of enactment of the Low-Level Radioactive Waste Policy Amendments Act of 1985, the Nuclear Regulatory Commission shall, in consultation with the States and other interested persons, identify methods for the disposal of low-level radioactive waste other than shallow land burial, and establish and publish technical guidance regarding licensing of facilities that use such methods.

"(b) Not later than 24 months after the date of enactment of the Low-Level Radioactive Waste Policy Amendments Act of 1985, the Commission shall, in consultation with the States and other interested persons, identify and publish all relevant technical information regarding the methods identified pursuant to subsection (a) that a State or compact must provide to the Commission in order to pursue such methods, together with the technical requirements that such facilities must meet, in the judgment of the Commission, if pursued as an alternative to shallow land burial. Such technical information and requirements shall include, but need not be limited to, site suitability, site design, facility operation, disposal site closure, and environmental monitoring, as necessary to meet the performance objectives established by the Commission for a licensed low-level radioactive waste disposal facility. The Commission shall specify and publish such requirements in a manner and form deemed appropriate by the Commission.

"SEC. 9. LICENSING REVIEW AND APPROVAL.

"In order to ensure the timely development of new low-level radioactive waste disposal facilities, the Nuclear Regulatory Commission or, as appropriate, agreement States, shall consider an application for a disposal facility license in accordance with the laws applicable to such application, except that the Commission and the agreement state shall—

"(1) not later than 12 months after the date of enactment of the Low-Level Radioactive Waste Policy Amendments Act of 1985, establish procedures and develop the technical capability for processing applications for such licenses;

"(2) to the extent practicable, complete all activities associated with the review and processing of any application for such a license (except for public hearings) no later than 15 months after the date of receipt of such application; and

"(3) to the extent practicable, consolidate all required technical and environmental reviews and public hearings.

"SEC. 10. RADIOACTIVE WASTE BELOW REGULATORY CONCERN.

"(a) Not later than 6 months after the date of enactment of the Low-Level Radioactive Waste Policy Amendments Act of 1985, the Commission shall establish standards and procedures, pursuant to existing authority, and develop the technical capability for considering and acting upon petitions to exempt specific radioactive waste streams from regulation by the Commission due to the presence of radionuclides in such waste streams in sufficiently low concentrations or quantities as to be below regulatory concern.

"(b) The standards and procedures established by the Commission pursuant to subsection (a) shall set forth all information required to be submitted to the Commission by licensees in support of such petitions, including, but not limited to—

"(1) a detailed description of the waste materials, including their origin, chemical composition, physical state, volume, and mass; and

"(2) the concentration or contamination levels, half-lives, and identities of the radionuclides present. Such standards and procedures shall provide that, upon receipt of a petition to exempt a specific radioactive waste stream from regulation by the Commission, the Commission shall determine in an expeditious manner whether the concentration or quantity of radionuclides present in such waste stream requires regulation by the Commission in order to protect the public health and safety. Where the Commission determines that regulation of a radioactive waste stream is not necessary to protect the public health and safety, the Commission shall take such steps as may be necessary, in an expeditious manner, to exempt the disposal of such radioactive waste from regulation by the Commission."

3. NEW YORK STATE LOW-LEVEL RADIOACTIVE WASTE MANAGEMENT ACT

CHAPTER 673

AN ACT to amend the public authorities law, the environmental conservation law, the public health law and chapter fifty of the laws of nineteen hundred eighty-six (STATE OPERATIONS BUDGET), in relation to the management of low-level radioactive waste generated in New York state

Became a law July 26, 1986, with the approval of the Governor. Passed on message of necessity pursuant to Article III, section 14 of the Constitution by a majority vote, three-fifths being present.

The People of the State of New York, represented in Senate and Assembly, do enact as follows:

Section 1. Short title. This act shall be known and may be cited as the "low-level radioactive waste management act".

§ 2. Legislative findings and declaration. The federal Low-Level Radioactive Waste Policy Act and the federal Low-Level Radioactive Waste Policy Amendments Act of 1985 make each state responsible for assuring adequate capacity for disposal of low-level radioactive waste generated within its borders. Due to the adoption of regional compacts pursuant to the federal acts and by the terms of these acts, New York cannot assume that other states will continue indefinitely to provide access to facil-

EXPLANATION—Matter in *italics* is new; matter in brackets [ ] is old law to be omitted.

ities adequate for the permanent disposal of low-level radioactive waste generated in New York. Pursuant to chapter nine hundred seventy-eight of the laws of nineteen hundred eighty-three, the state energy office has conducted a study of actions which should be taken by the state in response to the federal law and has issued a report which recommends that New York take immediate steps for siting and selection of disposal methods for construction and operation of facilities in New York for permanent disposal of low-level radioactive waste, and that the state undertakes to license, construct, and operate permanent disposal facilities pursuant to siting and disposal method selections by the department of environmental conservation. The legislature finds immediate implementation of steps toward establishing by January first, nineteen hundred ninety-three low-level radioactive waste management facilities, to be constructed and operated by the New York state energy research and development authority, acting through its own employees or, as provided through its existing general powers, through contractors, necessary to provide for continued operation of essential and benficial medical, research, industrial, energy and other facilities in New York which use radioactive materials and generate low-level radioactive waste and to protect the public health and safety and promote the general welfare of the people of the state of New York. The legislature further finds that, in these circumstances, the need to expedite the completion of these facilities and to minimize or eliminate any delay in beginning facility operation warrant the prescription of special provisions for the timely review and development of such facilities in a manner consistent with public participation in decision making, the protection of public health and safety and of the state's environment, and the promotion of the general welfare. The legislature further finds that the costs of siting, development, obtaining licenses, permits, or other approvals, construction, operation, maintenance, payments in lieu of taxes, contingencies, decontamination, decommissioning, closure, post-closure care, and all other costs and expenses of the state associated with low-level radioactive waste management facilities should be borne by the generators of such waste.

§ 3. Section eighteen hundred fifty-one of the public authorities law is amended by adding four new subdivisions fourteen, fifteen, sixteen and seventeen to read as follows:

14. "Low-level radioactive waste" shall mean radioactive waste that:

a. is not high-level radioactive waste, transuranic waste, spent nuclear fuel, or the tailings or wastes produced by the extraction or concentration of uranium or thorium from any ore processed primarily for its source material content; and

b. the United States nuclear regulatory commission, consistent with federal law, and in accordance with paragraph a of this subdivision, classifies as low-level radioactive waste.

15. "Low-level radioactive waste management facilities" shall mean facilities for permanent disposal of low-level radioactive waste and any associated facilities for treatment and handling of such waste, including but not limited to, facilities for purposes of stabilization, volume reduction, or the protection of health and safety of workers or members of the public.

16. "Permanent disposal facilities" shall mean low-level radioactive waste management facilities for permanent disposal of low-level radioactive waste generated within the state of New York other than such waste which is a federal responsibility pursuant to the provisions of federal law pertaining to state and federal responsibilities for disposal of low-level radioactive waste.

17. "Generate" or "generation", when used with respect to low-level radioactive waste, shall mean the production, or causing the production of, or activity which otherwise results in the creation or increase in volume of low-level radioactive waste. A person who generates low-level radioactive waste includes one who personally, or through the actions of any agent, employee, or contractor, generates low-level radioactive waste.

§ 4. Such law is amended by adding three new sections eighteen hundred fifty-four-b, eighteen hundred fifty-four-c and eighteen hundred fifty-four-d to read as follows:

§ *1854-b. Low-level radioactive waste management facilities. In addition to those purposes otherwise set forth in this title, the purposes of the authority shall include the management of low-level radioactive waste. In carrying out that purpose, the authority shall have the powers and responsibilities set forth in sections eighteen hundred fifty-four-c and eighteen hundred fifty-four-d of this title, in addition to those otherwise set forth in this title.*

§ *1854-c. Permanent disposal facilities. 1. The authority shall consult and cooperate with the department of environmental conservation and the commission for siting low-level radioactive waste disposal facilities pursuant to the provisions of article twenty-nine of the environmental conservation law and such department and commission shall make available to the authority all information with respect to potential sites and disposal methods which may be relevant to the design of, or preparation of applications for state licenses, permits, or other approvals for, such facilities. The authority shall make preparations for submitting as soon as practicable applications for any state licenses, permits, or other approvals required for the construction and operation of permanent disposal facilities.*

2. Upon certification by the department of environmental conservation of siting and disposal method selections for permanent disposal facilities pursuant to section 29-0105 of the environmental conservation law, the authority shall immediately complete development of and submit, as soon as practicable after such certification, but no later than the first day of January nineteen hundred ninety, complete applications for any state licenses, permits, or other approvals required for the construction and operation of such permanent disposal facilities, in accordance with the siting and disposal method certification of such department. Upon such certification, the authority shall also immediately take steps to acquire and hold in the name of the state of New York any property interests which it may need to obtain for such permanent disposal facilities.

3. Upon receipt of required state licenses, permits, or other approvals for such permanent disposal facilities and if access is not available on reasonable terms to disposal facilities outside of New York for low-level radioactive waste estimated to be generated within New York state during the life of such permanent disposal facilities, the authority is authorized and directed to immediately construct, and to operate and maintain, permanent disposal facilities in accordance with such licenses, permits, and other approvals. Permanent disposal facilities shall be completed and begin operation no later than the first day of January nineteen hundred ninety-three.

§ *1854-d. Generator reporting and fees. 1. Reports. a. Any person who generates low-level radioactive waste in New York shall submit to the authority, on dates specified by the authority, but in no event later than nine months after the effective date of the low-level radioactive waste management act and, thereafter, no less frequently than annually, reports detailing the classes and quantities of low-level radioactive waste generated, stored by the generator for decay or for later transfer to other facilities, or transferred by the generator to other facilities, the general type of generator (e.g., medical, university, industry, electric utility, government), and such additional information as the authority may reasonably require on the nature and characteristics (including, without limitation, chemical and physical characteristics, properties, or constituents, radionuclides present, curie content or concentration of radioactivity) of such waste and the extent of reduction in quantity and the nature and extent of reduction or other change in the nature or characteristics of such waste as a result of treatment or interim storage after generation and before delivery to facilities for permanent disposal of such waste. The authority shall provide by regulation appropriate procedures for the preparation and submission of such reports, including procedures to designate a person or persons responsible for such filing when more than one person is the generator of the same waste. Such reports shall be subject to the provisions of article six of the public officers law.*

b. Commencing no later than the first day of July nineteen hundred eighty-seven, the authority shall submit annually to the governor, the temporary president of the senate, the speaker of the assembly, the minority leader of the senate, and the minority leader of the assembly, and thereafter, not later than one hundred eighty days after the end of each calendar year, a report summarizing and categorizing, by type of generator and region of generation, the nature, characteristics, and quantities of low-level radioactive waste generated in New York during such calendar year.

EXPLANATION—Matter in *italics* is new; matter in brackets [ ] is old law to be omitted.

2. Fees. a. (i) Pursuant to this title the authority shall, pursuant to regulations promulgated in accordance with article two of the state administrative procedure act, establish, revise, assess, and collect reasonable rates, charges, or other fees upon the disposal of low-level radioactive waste generated in New York sufficient to fully recover all costs and expenses of the state and its agencies and authorities associated with low-level radioactive waste management facilities. Such assessed rates, charges, or other fees shall be paid to the authority in the manner, and accompanied by such returns, reports, or other documentation as the authority may prescribe. Fees paid shall be treated as expenses for purposes of recovery in rates. Surcharges collected by or for facilities which accept low-level radioactive waste generated in New York State for disposal pursuant to the Federal Low-level Radioactive Waste Policy Amendments of nineteen hundred eighty-five (Pub. Law 99-240), but deposited in escrow pursuant to section 5 (d) (2) of such law shall be paid to and received by the authority. Such payments shall be disbursed or transferred pursuant to this paragraph and shall be used for the purpose of reducing the amounts otherwise recoverable as assessments imposed pursuant to paragraph c of this subdivision.

(ii) The authority shall deposit the proceeds from such rates, charges, or other fees in a separate segregated account maintained solely for the purpose of holding such deposits. The authority shall first apply the proceeds of the account to payments of principal and interest on bonds, notes or other obligations issued by the authority for the purposes of the low-level radioactive waste management act. Upon appropriation or pursuant to other legislative authorization, disbursements or transfers from such account shall be made solely for other costs and expenses incurred under the low-level radioactive waste management act.

(iii) In establishing, revising, assessing, and collecting such reasonable rates, charges, or other fees associated with low-level radioactive waste management facilities or incurred under the low-level radioactive waste management act, the authority shall include, but need not be limited to, such amounts as will meet all incremental costs and expenses of:

A. selecting, developing, licensing, constructing, operating, and maintaining such facilities;

B. establishing reserves for related purposes, including, but not limited to, debt service, decontamination and decommissioning, closure, and post-closure care, and contingencies;

C. making reimbursements of expenditures from appropriations to state agencies or authorities to implement the low-level radioactive waste management act; and

D. making payments in lieu of taxes.

(iv) The authority may establish and revise, not inconsistent with federal law and regulation, reasonable classifications of low-level radioactive waste based upon the nature, class, characteristics (other than type of generator) or condition of such waste. The authority may utilize such classifications for purposes of establishing or revising the rates, charges, or other fees upon the disposal of such waste pursuant to paragraphs b and c of this subdivision.

b. Upon delivery of low-level radioactive waste to permanent disposal facilities owned or operated by the authority, the rates, charges, or other fees as established or revised by the authority for disposal of such waste at such facilities, and related services, shall be paid. Such rates, charges, or other fees shall be sufficient to meet all costs and expenses which may be reasonably and practically identified or allocated to permanent disposal facilities, over the life of such facilities, and the low-level radioactive waste disposed of at such facilities, including without limitation, all costs and expenses associated with: development, licensing, operation, maintenance, and meeting debt service requirements, or requirements for repayment of expenditures from appropriations for capital costs of such facilities, which requirements are due after such facilities begin accepting low-level radioactive waste; establishing reserves, including but not limited to reserves for decontamination and decommissioning, closure and post-closure care, and contingencies, including but not limited to potential accidents, damages, and injuries; and making any payments in lieu of taxes or fees for local assistance and repayments of any amounts expended from appropriations for aid to localities with respect to permanent disposal facilities, provided, however, that all capital costs incurred prior to the receipt and acceptance of low-level radioactive waste at the disposal facilities shall be recovered in a period of not less than twenty years after such receipt and acceptance.

c. During the period before the commencement of operation of permanent disposal facilities owned or operated by the authority, expenditures from appropriations to the departments of environmental conservation and health and the commission for siting low-level radioactive waste disposal facilities for the purpose of implementing the low-level radioactive waste management act shall be recovered through assessment against United States nuclear regulatory commission licensees for nuclear electric generating facilities located in the state of New York which have full power operating licenses. For the state fiscal year ending March thirty-first, nineteen hundred eighty-seven and each successive state fiscal year until commencement of operation of the permanent disposal facilities, the chairman of the authority shall estimate such expenditures and shall assess such estimated expenditures on a pro rata basis in proportion to the number of reactors with full power operating licenses of each such licensee. For the assessment for the state fiscal year ending March thirty-first, nineteen hundred eighty-seven, such estimated expenditures shall be based upon actual appropriations available to such state agencies for such fiscal year for such purpose and shall be billed by the authority as soon as practicable after the effective date of the low-level radioactive waste management act and paid by such licensees not later than December thirty-first, nineteen hundred eighty-six. For the assessment for each subsequent fiscal year, such estimated expenditures shall be based upon the proposed appropriations to such state agencies for such purpose, net of any over or under assessment for prior fiscal years compared to the latest available data on such expenditures for such prior fiscal years, and shall be billed by the authority on or before February first preceding such fiscal year and paid on or before April first of such fiscal year; provided, however, that a licensee may elect to make partial payments for such assessments on March tenth of the preceding fiscal year and on June tenth, September tenth, and December tenth of such fiscal year. Each such partial payment shall be an amount not less than twenty-five per centum of the total annual assessment against such licensee for the relevant fiscal year. The authority shall establish and maintain records to account for assessments made against and received from each licensee. Upon receipt of such assessments, and pursuant to an agreement with the director of the budget, the authority shall transmit to the state comptroller such amount as shall be required to reimburse actual expenditures made and subject to assessment. The amounts so assessed, notwithstanding their assessment, shall be included as cost and expenses for purposes of computing and imposing rates, charges, or other fees pursuant to paragraphs a and b of this subdivision. In imposing such rates, charges, or other fees pursuant to such paragraph b, the authority shall provide a credit to the assessed licensees until such time as the aggregate of all such credits for a licensee shall equal the actual assessments paid by such licensee, plus interest at a reasonable market rate.

d. (i) The authority may refuse to accept at facilities established pursuant to section eighteen hundred fifty-four-b of this title any low-level radioactive waste generated by a person who has failed to submit to the authority, the one or more reports required by the authority pursuant to subdivision one of this section.

(ii) The proceeds of any penalty or interest collected pursuant to subparagraph (i) of paragraph a of subdivision three of this section shall be remitted to the authority for the purposes of the low-level radioactive waste management act.

3. Violations. a. Any failure or refusal to file a report, return, or other documentation, or related information, required pursuant to the provisions of this section shall be deemed a violation of the provisions of, and a failure to perform a duty imposed by, this section and shall be subject to the following civil and criminal penalties:

(i) By a civil penalty, in the case of a first violation, not to exceed five thousand dollars, and in the case of a second or subsequent violation, a civil penalty not to exceed ten thousand dollars; which penalty may be assessed and collected by a court in any action or proceeding pursuant to subparagraph (ii) of this paragraph in addition to any criminal penalty which may be assessed for such violation.

EXPLANATION—Matter in *italics* is new; matter in brackets [ ] is old law to be omitted.

(ii) By a misdemeanor, in the case of a willful violation by a person having any of the culpable mental states defined in section 15.05 of the penal law, which shall be deemed a misdemeanor, and upon a first conviction thereof, by a fine not to exceed five thousand dollars, or by imprisonment for a term of not more than six months, or both such fine and imprisonment; and, upon a second or subsequent conviction thereof, punishment by a fine not to exceed ten thousand dollars, or by imprisonment for a term of not more than one year, or by both such fine and imprisonment.

b. The attorney general shall institute such civil proceedings as the authority may request for the purpose of enforcing the provisions of this section, and such criminal proceedings as the authority may request for the purpose of prosecuting criminal violations of this section.

4. Upon appropriation or other legislative authorization and consistent with a repayment agreement executed with the director of the budget, the authority shall repay from the special account established pursuant to subparagraph (ii) of paragraph a of subdivision two of this section to the general fund, capital projects fund, or any other fund all amounts expended from appropriations made to the authority, the department of environmental conservation, the commission for siting low-level radioactive waste disposal facilities, the department of health, or the department of labor for actual and incremental expenses in the selection, development, licensing, construction, operation, maintenance, decontamination and decommissioning, closure and post-closure care of low-level radioactive waste management facilities, including any regulatory program associated therewith, or for aid to localities.

5. The authority may establish reasonable terms and conditions for receipt, acceptance or disposal of low-level radioactive waste at any facilities developed pursuant to section eighteen hundred fifty-four-c of this title, including but not limited to packaging, identification of the nature and sources of the waste and the generators, shippers, or carriers of such waste, or other persons having any care or custody of such waste from the place of generation to arrival at such facilities, and requirements for bonds, insurance, or other forms of financial protection or assurance of performance by persons responsible for such waste.

6. Title to any low-level radioactive waste shall vest in the state of New York upon acceptance of such waste by the authority at the permanent disposal facilities. Acceptance at permanent disposal facilities shall not occur until completion of such inspection and examination of the waste and determination of compliance with applicable terms and conditions, including but not limited to payment of applicable fees, as the authority may require.

7. Notwithstanding the provisions of subdivision twelve of section eighteen hundred fifty-five of this title, the authority shall enter into agreements to pay annual sums in lieu of taxes to any municipality or taxing district of the state with respect to any real property acquired and held by the authority in the state or improvements to any real property held by the authority in the state, which real property was acquired or improvements were made after the effective date of the low-level radioactive waste management act and for the purposes of such act; provided, that any amount so paid shall be recovered through rates, charges, or other fees applicable to disposal of low-level radioactive waste, pursuant to subdivision two of this section.

8. Actions by the authority pursuant to the provisions of section eighteen hundred fifty-four-c of this title shall be subject, where applicable, to the environmental and judicial review provisions set forth in article twenty-nine of the environmental conservation law.

§ 5. The environmental conservation law is amended by adding a new article twenty-nine to read as follows:

ARTICLE 29

LOW-LEVEL RADIOACTIVE WASTE FACILITIES

TITLE 1. Siting.
3. Commission for siting low-level radioactive waste disposal facilities.
5. Advisory committee on permanent disposal facilities siting and disposal method selection.
7. Financial assurance.

TITLE 1

SITING

Section 29-0101. Definitions.
29-0103. Siting criteria for permanent disposal facilities.
29-0105. Certification of site and disposal method selection.

§ 29-0101. Definitions.

For the purposes of this article:

1. "Low-level radioactive waste" means radioactive material that:

a. is not high-level radioactive waste, transuranic waste, spent nuclear fuel, or the tailings or wastes produced by the extraction or concentration of uranium or thorium from any ore processed primarily for its source material content; and

b. the United States nuclear regulatory commission, consistent with federal law and in accordance with paragraph a of this subdivision, classifies as low-level radioactive waste.

2. "Low-level radioactive waste management facilities" means facilities authorized pursuant to section eighteen hundred fifty-four-c of the public authorities law for permanent disposal of low-level radioactive waste and any associated facilities for treatment and handling of such waste, including, but not limited, to facilities for purposes of stabilization, volume reduction, or protection of health and safety of workers or members of the public from potential exposure to hazards.

3. "Permanent disposal facilities" means low-level radioactive waste management facilities for permanent disposal of low-level radioactive waste generated within the state of New York, other than such waste which is a federal responsibility pursuant to the provisions of federal law pertaining to state and federal responsibilities for disposal of low-level radioactive waste.

4. "Shallow land burial" means emplacement of low-level radioactive waste in or within the upper thirty meters of the surface of the earth in trenches, holes, or other excavations in which only soil provides structural integrity, a barrier to migration of low-level radioactive waste from or subsurface water into such excavation, or a barrier to entry of surface water to such excavation or in a manner that fails to allow during the institutional control period for monitoring and control of releases of radioactivity.

5. "State agency" means any office, department, board, commission, bureau, division, council, authority, corporation, agency, or instrumentality of the state.

6. "Commission" means the commission for siting low-level radioactive waste disposal facilities created pursuant to section 29-0301 of this article.

§ 29-0103. Siting criteria for permanent disposal facilities.

1. No later than July thirty-first, nineteen hundred eighty-seven, the department shall publish draft regulations which specify the criteria for siting permanent disposal facilities and shall promulgate final regulations no later than December thirty-first, nineteen hundred eighty-seven. Such regulations shall be specific to the types of disposal methods which may be employed at a permanent disposal site and shall include criteria for:

a. above ground, engineered, monitored disposal;

b. underground mined repository disposal; and

c. where practicable, other disposal methods for which there are applicable regulations but in no event including shallow land burial.

Such regulations shall specify the minimum characteristics a disposal site and a disposal method must have under existing federal and state law to be acceptable for use for permanent disposal facilities.

2. The department shall hold hearings on the proposed siting criteria regulations.

3. In adopting the siting criteria regulations the department shall not be subject to the requirements of sections two hundred two-a, two hundred two-b and two hundred two-c of the state administrative procedure act.

4. The siting criteria regulations shall not be subject to the review and approval of the environmental board.

§ 29-0105. Certification of site and disposal method selection.

EXPLANATION—Matter in *italics* is new; matter in brackets [ ] is old law to be omitted.

1. Upon application by the commission, the department shall certify that the commission's selection of the site or sites and disposal method or methods pursuant to section 29-0305 of this article is in conformance with the applicable siting criteria promulgated pursuant to section 29-0103 of this title; shall certify that, with modifications proposed by the department, the selection would be in conformance with the applicable siting criteria; or shall refuse to certify such selection and shall specify the manner in which such selection fails to meet the applicable siting criteria. Such certification or refusal to certify shall be based upon the record. The department shall hold public hearings with respect to the draft environmental impact statement and with respect to the application for certification. Such hearings and such certification or refusal to certify shall be completed within one hundred eighty days of the commission's submission of a complete application for certification.

2. The department shall publish in the state register notice of its decision on the application for certification of the site and disposal method selection.

3. The department's decision on the application for certification of the site and disposal method selection and the accompanying final environmental impact statement shall be submitted by the department to the governor, the legislature, and the chairman of the New York state energy research and development authority. In addition to any other information otherwise required for a final environmental impact statement, such statement shall include:

a. copies of the minutes of the public hearing held on the draft environmental impact statement on siting and disposal method selection and of recommendations from the advisory committee established pursuant to section 29-0501 of this article, and the department's responses to the views, comments, information, and recommendations therein; and

b. a listing providing a brief description, identification, or reference for each report, study, or other document relied upon by the department for information supporting its analyses or conclusions.

TITLE 3

COMMISSION FOR SITING LOW-LEVEL RADIOACTIVE WASTE DISPOSAL FACILITIES

§ 29-0301. Appointment of the commission.

1. There is hereby created a commission for siting low-level radioactive waste disposal facilities. The commission shall consist of five members appointed by the governor whose appointment shall be effective when issued. The members of the commission, to the extent practicable, shall be competent and knowledgeable concerning low-level radioactive waste and shall be qualified as follows:

a. a geologist;
b. a medical doctor;
c. a health physicist;
d. a professional engineer;
e. a private citizen who shall be designated to act as chairperson.

2. The chairperson shall appoint an executive director to the commission.

3. In the event of resignation of one of the commission members, the governor shall appoint a replacement. Each member shall receive the sum of two hundred dollars for each day in which the member is actually and primarily engaged in the performance of the duties specified herein plus actual and necessary expenses incurred by such member in the performance of such duties.

4. The members of the commission, the executive director and any employees of the commission shall be considered public officers for purposes of the public officers law.

5. Three of the five members of the commission shall constitute a quorum for the transaction of business of the commission and the decision of three members of the commission shall constitute action of the commission; provided, however, that no application for certification pursuant to section 29-0105 of this article shall be made except upon the affirmative vote of three or more members of the commission.

§ 29-0303. Duties of the commission.

1. The commission shall immediately commence the preparation of a siting and disposal method selection which shall, upon certification by the department, be the site or sites and method or methods for permanent disposal facilities which shall be constructed or operated by the energy research and development authority pursuant to section eighteen hundred fifty-four-c of the public authorities law.

2. No later than the first day of December, nineteen hundred eighty-eight, the commission shall complete the preparation of its site and disposal method selection and shall submit its application for certification of this selection by the department pursuant to section 29-0105 of this article. The commission shall simultaneously deliver a copy of its draft environmental impact statement and application to the governor, the legislature and the chairman of the energy research and development authority.

3. The site and disposal method selection and the application for certification, together with the department's certification thereof, shall be considered a single action for purposes of article eight of this chapter and judicial review. The commission shall prepare a draft environmental impact statement to accompany its application for certification, the scope of which must be approved by the department. For purposes of satisfying the requirements of article eight of this chapter, the department shall be the lead agency.

4. The commission's site and disposal method selection shall identify a site or sites and appropriate disposal method or methods for permanent disposal facilities. Such site or sites shall not include the western New York nuclear service center. The commission shall take into account the following factors in the selection of the permanent disposal facility site or sites and disposal method or methods:

a. the nature and probability of the impacts on public health and safety, including predictable adverse effects from:

(i) accidents during transportation of low-level radioactive waste to such facilities;

(ii) contamination of ground or surface water by leaching and runoff from such facilities; and

(iii) fires or explosions from improper storage or disposal of volatile, combustible, or potentially explosive materials, if any, which may compose a portion of the low-level radioactive waste to be delivered to such facilities;

b. the nature of the probable environmental impacts, including the predictable adverse effects on the natural environment and ecology, scenic, historic, cultural, and recreational values, water and air quality, and wildlife;

c. the potential for avoidance or mitigation of harm from the unanticipated release of low-level radioactive waste or contaminated materials;

d. the ability for retrieval or recovery of such waste;

e. differences in the density of population in the vicinity of the potential sites;

f. the adequacy of routes and means for transportation of low-level radioactive waste to such facilities;

g. the nature of the probable impact of such facilities on local governmental units within which such facilities would be located; and

h. the comparative economic implications, including those resulting from engineering considerations, of the potential site or sites and disposal methods for such facilities.

5. The commission shall select one site for a permanent disposal facility after consideration of all relevant public health and safety, environmental and economic factors, provided, however, that an additional site may be selected if the commission finds that the use of an additional site presents specific advantages with respect to such factors. To the extent the commission determines that different disposal methods are appropriate for different categories of low-level radioactive waste with differing physical or chemical characteristics, the commission may select more than one disposal method to be utilized at each particular site, specifying the particular disposal methods to be uti-

EXPLANATION—Matter in *italics* is new; matter in brackets [ ] is old law to be omitted.

lized at such site for particular categories of such waste; provided that utilization of the disposal methods selected at the site selected shall be capable of meeting or exceeding applicable requirements of state and federal regulations. The site or sites selected shall be of sufficient capacity to provide for disposal, using the selected disposal methods, of all low-level radioactive waste estimated by the commission to be generated in New York and to require disposal at low-level radioactive waste management facilities for a period of at least thirty years.

6. In performing its duties, the commission shall hold periodic meetings which shall be publicly noticed pursuant to article seven of the public officers law.

§ 29-0305. Operation of the commission.

1. The commission may hire or contract with such persons as it deems necessary, convenient, or desirable to carry out its duties pursuant to this title.

2. Officers and employees of the commission shall be appointed in accordance with civil service rules; provided, however, that officers and employees of state departments and agencies may be transferred to the commission without examination and without loss of any civil service status or rights. Each employee who is transferred pursuant to this subdivision is deemed to be on leave of absence from his or her former position during the tenure of the commission. No such transfer may, however, be made except with the approval of the head of the state department or agency involved, the director of the budget and the chairman of the commission, and in compliance with the rules and regulations of the civil service commission of the state.

3. The commission, its employees, agents, consultants, or contractors may, after proper notification and identification, enter at reasonable times upon such lands, waters or premises as in the judgment of the commission may be necessary, convenient, or desirable for the purpose of making surveys, soundings, borings, and examinations to accomplish any purposes authorized by this title, the commission being liable for actual damage done. Each such entry shall be commenced and completed with reasonable promptness. If the officer or employee obtains any samples prior to his leaving the premises, he shall give to the owner, operator, or agent in charge a receipt describing the sample obtained and, if requested, a portion of each such sample equal in volume or weight to the portion retained. If any analysis is made of such samples, a copy of the results of such analysis shall be furnished promptly to the owner, operator, or agent in charge.

§ 29-0307. Cooperation with other agencies.

1. All agencies and authorities of the state, municipalities, and political subdivisions of the state are hereby directed to cooperate with the commission in order to facilitate and expedite the responsibilities and duties of the commission. To the extent practicable and not otherwise inconsistent with law and upon the request of the commission, any agency or authority of the state shall provide copies of existing studies, surveys, plans, data, and other materials in its possession to the commission.

2. The commission shall assure that the advisory committee appointed pursuant to section 29-0501 of this article has a timely opportunity to provide information and recommendations to the commission on all aspects of its activities pursuant to section 29-0303 of this title.

3. In consultation and cooperation with the advisory committee established pursuant to section 29-0501 of this article, the commission shall keep the public informed of its activities in developing the draft environmental impact statement required by section 29-0303 of this title and encourage the public to participate by providing views, comments, information, and analysis concerning siting and disposal method selection for permanent disposal facilities.

4. The department of audit and control and the department of law are hereby directed to expedite the processing of all contracts associated with carrying out the provisions of this section.

§ 29-0309. Tenure of the commission.

The commission shall continue in existence only until final judicial review of the department's certification of the commission's selection pursuant to section 29-0105 of this article.

TITLE 5

ADVISORY COMMITTEE ON PERMANENT DISPOSAL FACILITIES SITING AND DISPOSAL METHOD SELECTION

§ 29-0501. Advisory committee on permanent disposal facilities siting and disposal method selection.

1. An advisory committee on siting and disposal method selection for permanent disposal facilities is hereby established within the commission and shall continue in existence until the department has issued its final environmental impact statement pursuant to section 29-0105 of this article. The members shall be appointed by the governor and shall consist of: the state geologist; the commissioners of health, labor, the state energy office, and transportation and the secretary of state, or their respective designees; two representatives of non-profit environmental organizations; two health physicists or medical doctors knowledgable of radiation health effects; two representatives of low-level radioactive waste generators in New York; a private citizen residing in New York who is technically competent and knowledgeable concerning low-level radioactive waste; and, upon issuance of the commission's draft environmental impact statement pursuant to section 29-0303 of this article, three private citizens from the county within which each proposed site is primarily located. The governor shall designate the chairman from among the members of the committee not serving in an ex officio capacity. The chairman of the committee shall designate a person to serve as executive secretary to the advisory committee. The chairman of the commission shall make staff of the commission available to support the committee in carrying out its responsibilities. No member shall receive any compensation, but shall be entitled to reimbursement by the commission for actual and necessary expenses in performing the duties of the advisory committee.*

2. The advisory committee shall:

a. meet at least bi-monthly to provide information and review and assist activities of the commission and the department pursuant to sections 29-0103, 29-0105, 29-0303 and 29-0509 of this article and to receive a written report from the commission, the department and the energy research and development authority on plans and progress in carrying out activities and duties pursuant to the low-level radioactive waste management act. In particular, the advisory committee shall provide information and recommendations in response to the commission's draft environmental impact statement filed pursuant to subdivision three of section 29-0103 of this article, including reviewing public views, comments, and information submitted in response thereto;

b. review and comment semi-annually on the commission's plans and schedule for carrying out the provisions of section 29-0303 of this article;

c. have minutes taken of each meeting of the advisory committee and make them available to the public within ten working days from the date of the meeting;

d. consult with and advise the commissioner of health in planning and carrying out a public information and education program on low-level radioactive waste pursuant to the provisions of article twenty-four-C of the public health law; and

e. exercise and perform such other advisory functions related to the commission's or the department's activities conducted pursuant to sections 29-0103, 29-0105, 29-0303 and 29-0509 of this article as the

---

* So in original. (Word misspelled.)

EXPLANATION—Matter in *italics* is new; matter in brackets [ ] is old law to be omitted.

chairman of the commission or commissioner of environmental conservation may request.

3. All agencies and authorities of the state are hereby directed to cooperate with the advisory committee in order to facilitate and expedite the responsibilities and duties of the commission.

§ 29-0503. State agency actions on licenses, permits, or approvals for low-level radioactive waste management facilities.

1. With respect to any particular permanent disposal facilities, all applications for state licenses, permits, or other approvals required for those facilities shall be submitted contemporaneously to the respective state agencies with jurisdiction to grant such licenses, permits, or other approvals; and shall be accompanied by a draft environmental impact statement for those facilities and a list identifying each state license, permit, or other approval for which such applications have been submitted and the jurisdictional state agency for such license, permit, or other approval.

2. Notwithstanding any other provision of law, all applications to a single state agency for required state licenses, permits, or other approvals for particular low-level radioactive waste management facilities shall be consolidated by such state agency and considered in a single proceeding, which shall be completed as expeditiously as possible.

3. All state agencies to which applications for required licenses, permits, or other approvals for particular low-level radioactive waste management facilities have been submitted shall keep each other informed of the procedural status of such applications and the proceedings thereon.

4. With respect to the proceedings on applications for required state licenses, permits, and other approvals for particular low-level radioactive waste management facilities:

a. If any such license, permit, or other approval for the particular low-level radioactive waste management facilities in question is within the jurisdiction of the department, the department shall be the lead agency with respect to environmental review of all applications to state agencies for such licenses, permits, or other approvals.

b. No later than thirty days after submission to the lead agency and other state agencies of such applications, the lead agency and each such other state agency shall give notice to the applicant that such applications within their respective jurisdictions have been determined to be complete or have been determined to be incomplete; provided, however, that when there is a requirement pursuant to federal law for a tentative determination or draft permit to be prepared prior to public notice or hearing, the time within which the agency shall make its determination whether or not the application is complete shall be extended by thirty days. If any such application has been determined to be incomplete, such notice shall include a detailed list of specific deficiencies in such application.

c. No later than sixty days after the lead agency and other jurisdictional state agencies have made their respective determinations that such applications within their respective jurisdictions are complete, the lead agency shall begin public hearings on the draft environmental impact statement and all other matters related to such applications. Any state agency, other than the lead agency, which determines to conduct public hearings with respect to any action or proceeding before it on such applications shall conduct such public hearings jointly with the public hearings conducted by the lead agency with respect to such facilities. The department shall hold an issues conference prior to the commencement of the hearing. At least one hearing shall be held at a reasonably convenient location in the general geographic vicinity of each of the proposed sites.

d. No later than one hundred fifty days after the commencement of such hearings for any low-level radioactive waste management facilities, such hearings and the period for the receipt of any written comments, arguments, or analyses with respect to matters raised in such hearings shall have been completed.

e. No later than ninety days after completion of such hearings and the period for the receipt of written comments, arguments, or analyses with respect to matters raised in such hearings, the lead agency shall issue a final environmental impact statement related to the applications which were the subject of such hearings. In addition to any other information otherwise required for a final environmental impact statement, such statement shall include:

(i) Copies of the minutes of the public hearings held on the draft environmental impact statement associated with a state agency action on a license, permit, or approval for a low-level radioactive waste management facility, and the department's responses to the views, comments, information and recommendations thereon; and

(ii) A listing providing a brief description, identification, or reference for each report, study, or other document relied upon by the department for information supporting its analyses or conclusions.

f. The lead agency shall keep each other state agency before which any such application is pending informed of the progress of its development of the final environmental impact statement. Immediately upon issuance of the final environmental impact statement, the lead agency shall deliver a copy to each such other state agency. No later than thirty days after the issuance of such final environmental impact statement, the lead agency and each such other state agency shall issue their decisions with respect to such licenses, permits, and other approvals with any reasonable modifications or conditions which the lead agency, and each such other state agency, respectively, finds required in accordance with the provisions of law and regulations applicable to its respective action or proceeding. Each agency shall publish notice in the state register of its decision with respect to such licensing or other approval. Each such decision shall be based upon the administrative record for the respective action or proceeding.

5. In any action or proceeding of the department or any other state agency on any application for a required state license, permit, or other approval for any low-level radioactive waste management facilities, including any related draft or final environmental impact statement proposed or submitted in connection with such application, the following matters as determined by statute or certified pursuant to section 29-0105 of this article shall not be in issue:

a. the need for such facilities or the alternative of no action;

b. the site or sites of such facilities;

c. for permanent disposal facilities, the disposal methods to be utilized;

d. the nature or type of facilities as specifically required or authorized by statute; and

e. the classes of waste which may be stored or disposed of at such facilities.

§ *29-0505. General provisions on environmental review and judicial review.*

Notwithstanding any other provision of law:

1. In the event of any inconsistency between the provisions of this article and the provisions of article eight or seventy of this chapter, or regulations issued pursuant thereto, the provisions of this article shall have precedence and apply to the exclusion of such provisions of article eight or seventy of this chapter or the regulations issued pursuant thereto.

2. a. Any person aggrieved by any administrative action or proceeding in connection with the adoption of siting criteria regulations pursuant to section 29-0103 of this article, with a certification or refusal to certify a site or disposal method selection pursuant to section 29-0105 of this article or with the issuance or denial of a required state license, permit, or other approval for low-level radioactive waste management facilities may seek judicial review of such administrative action or proceeding in accordance with the provisions of this subdivision. Any such special proceeding for judicial review shall be brought in the appellate division of the supreme court of the judicial department embracing the county wherein the site of the facilities is located, or, if the certification or the application for a state license, permit, or other approval is denied, the county wherein the commission or applicant proposed to site or locate the facilities. Such review may be initiated only by the filing of a petition in such court within thirty days after publication in the state register of notice of the administrative action or decision, together with proof of service of a demand on the commission, the department, and other state agencies, as applicable, for the filing with the court of a copy of the administrative record. Upon receipt of such petition and demand, a copy of the administrative record and any decision shall forthwith be delivered by the commission, the department, or other state agency, as applicable, to the court. The pe-

EXPLANATION—Matter in *italics* is new; matter in brackets [ ] is old law to be omitted.

tition and any subsequent appeal shall be heard on the administrative record without requirement of reproduction. No objection that has not been urged on the administrative record shall be considered by the court, unless the failure or neglect to urge such objection below shall be excused because the information underlying such objection was unknown at the time of the administrative proceeding or because of other extraordinary circumstances.

b. When judicial review is sought of the department's siting criteria regulations or the department's certification of the commission's siting and disposal method selection, neither the appellate division nor the court of appeals shall have any jurisdiction or power to issue any temporary or preliminary order staying, pending completion of judicial review, such selection or certification or staying, pending completion of judicial review, any subsequent proceeding on an application for any required state license, permit, or other approval for low-level radioactive waste management facilities based upon the siting and disposal method selection certified by the department pursuant to section 29-0105 of this article.

c. If such facilities are proposed to be sited or located in more than one judicial department, such proceeding may be brought in any one but only one of such departments. If petitions are filed in more than one court, the court in which a petition was first filed shall retain exclusive jurisdiction of the proceeding, and all other petitions shall be transferred forthwith to said court. Upon motion by any party to the proceeding, or on its own motion, said court may transfer the proceedings to the appellate division in any other judicial department for good cause. The jurisdiction of the appellate division shall be exclusive and its judgment and order shall be final, subject to review by the court of appeals in the same manner and form and with the same effect as provided for appeals in a special proceeding. All such special proceedings shall be heard and determined by the appellate division and by the court of appeals as expeditiously as possible and with precedence over all other matters except special proceedings under the election law.

d. Except as otherwise provided in this subdivision, article seventy-eight of the civil practice law and rules shall apply to special proceedings and appeals therefrom taken pursuant to this subdivision.

§ 29-0507. State licenses, permits, and other approvals exclusive.

Notwithstanding any other provision of law, no county, city, village, town, or special district, or any agency or instrumentality thereof, may prohibit, or require any license, permit, other approval, or condition related to, the construction or operation of low-level radioactive waste management facilities for which required state licenses, permits, or other approvals have been issued. Nothing in this section shall preclude any county, city, village, town, or special district, or any agency or instrumentality thereof, from requesting, proposing, or advocating imposition of reasonable conditions on the construction or operation of low-level radioactive waste management facilities in any proceeding upon an application for a required state license, permit, or other approval for such facilities.

§ 29-0509. Aid to local governments.

No later than the first day of April, nineteen hundred eighty-seven, the department shall submit to the legislature and the governor a report recommending appropriate forms of state aid to local governmental units within the boundaries of which any low-level radioactive waste management facilities may be located. Such report shall be developed by the department and shall:

1. describe the nature of the probable impacts upon local governmental units, including both impacts normally expected from the presence, construction, operation, maintenance, closure, and post-closure care of such low-level radioactive waste management facilities and impacts which might result from an emergency or other abnormal or unusual event associated with the operation, maintenance, closure, or post-closure care of such low-level radioactive waste facilities;

2. describe the possible forms or kinds of aid or assistance which might be appropriate to mitigate or provide off-setting benefits with respect to each of the kinds of probable impacts on local governmental units identified pursuant to this section; and

3. set forth the department's specific recommendations to the legislature and the governor for the forms or kinds of state assistance to local governmental units to mitigate or provide off-setting benefits with respect to such impacts, together with the reasons for the department's specific recommendations.

TITLE 7

FINANCIAL ASSURANCE

Section 29-0701. *Financial requirements for low-level radioactive waste disposal facilities.*
29-0703. *Closure and post-closure plans.*

§ 29-0701. *Financial requirements for low-level radioactive waste disposal facilities.*

1. Within eighteen months after the effective date of this section, the commissioner shall promulgate regulations applicable to facilities for the permanent disposal of low-level radioactive waste which identify financial requirements to be included as conditions in permits for facilities for the management of low-level radioactive waste. Such conditions shall provide for the remediation of failures during operation and after facility closure, for facility closure, and for pre-closure and post-closure facility monitoring and maintenance. Such regulations shall:

a. Reflect due consideration of the sizes and locations of affected facilities, the natures and volume of low-level radioactive waste involved, the types of facilities and the degrees and durations of risk to human health or the environment.

b. Provide for the establishment, administration, terms and conditions of the following methods or instruments to be used as alternatives or in combinations, in order to achieve non-duplicative coverage of the financial assurance requirements mandated by this section:

(i) Trust funds.

(ii) Surety or performance bonds.

(iii) Letters of credit.

(iv) Liability insurance or annuities.

(v) Guarantees provided by corporate or other legal or financial affiliates of the facility owner or operator.

c. Establish the duration of such financial requirements.

2. Any owner or operator of such a facility for the management of low-level radioactive waste may request a modification of any of the financial requirements established pursuant to subdivision one of this section. A modification may be granted in the discretion of the department if such financial requirements are found to be unnecessary or inappropriate, consistent with the public interest and the purposes of this section and supported by written findings setting forth the reasons for the modification. Such a modification request shall be considered a request for modification of the permit for the facility. In no case shall a modification granted pursuant to this subdivision eliminate or reduce the minimum requirements established in subdivision four of this section.

3. In addition to the financial requirements established pursuant to subdivision one of this section, any permits for such facilities for the management of low-level radioactive waste issued by the department may, if it is determined that adequate protection of the public so requires, include conditions related to any or all of the following, including responsibility for the costs thereof:

a. On-site environmental monitors whose function shall be to monitor compliance with permit conditions. The commissioner may promulgate regulations regarding the use of such monitors.

b. Site safety plans whereby the permittee shall establish, with the cooperation of local government officials, a community and project safety plan, including but not limited to an accident response based on a worst-case condition, on-site and off-site, a personnel training program, provisions for coordination with local emergency services and regular training exercises. Any such plan shall be subject to the approval of the department.

4. Any permits for such facilities for the management of low-level radioactive waste issued by the department shall require the owner or operator to provide, at a minimum, one of the methods or instruments of financial assurance provided for in paragraph b of subdivision one of this section. Such methods or instruments shall be designed to insure proper facility closure, based on the estimates approved pursuant to section 29-0703 of this title, and coverage of personal injury and

EXPLANATION—Matter in *italics* is new; matter in brackets [ ] is old law to be omitted.

property damage to third parties caused by the operation of such facility. Such methods or instruments shall from time to time be reviewed and updated, pursuant to regulations promulgated by the department, to insure their continued adequacy for the purposes of this section.

§ 29-0703. Closure and post-closure plans.

1. Any permits for such facilities for the management of low-level radioactive waste issued by the department shall require owners and operators of facilities for the permanent disposal of low-level radioactive waste to submit to the department for its approval plans for the closure and post-closure monitoring and maintenance of their facilities. The department may promulgate rules and regulations concerning the contents of such plans. Such plans shall be approved prior to the effective date of the facilities' operating permit.

2. Together with the submission of a plan for closure and post-closure monitoring and maintenance, the owner or operator of such facilities for the permanent disposal of low-level radioactive waste shall submit to the department for approval a written estimate of the costs associated with closure at the estimated point in the facilities' operating life when the extent and manner of their operation would make closure the most expensive, as indicated by the closure plan.

3. The owner or operator shall prepare new closure and post-closure monitoring and maintenance cost estimates whenever a change in the closure and post-closure monitoring and maintenance requirements affects the cost of the closure or post-closure care.

4. One year after the approval of the cost estimates as set forth in subdivision two of this section, and annually thereafter until closure of the facilities, the operator's closure and post-closure monitoring and maintenance estimates shall be adjusted to account for inflation.

5. All estimates made pursuant to this section and revision thereto shall be subject to the review and approval of the department.

6. The methods and instruments of financial assurances for permitted facilities for permanent disposal of low-level radioactive waste shall be periodically reviewed by the department to determine whether they are adequate in light of changed circumstances to insure proper closure and post-closure monitoring and maintenance of such facilities.

§ 6. The public health law is amended by adding a new article twenty-four-C to read as follows:

ARTICLE 24-C

INFORMATION PROGRAM ON LOW-LEVEL RADIOACTIVE WASTE

Section 2485. Program to inform and educate the public on low-level radioactive waste.

§ 2485. Program to inform and educate the public on low-level radioactive waste. In consultation and cooperation with the advisory committee established pursuant to section 29-0501 of the environmental conservation law, and with the department of environmental conservation, the commissioner of health shall plan and carry out a statewide public information program on the public health and safety implications of low-level radioactive waste management. The content of such program shall include a basic explanation of the types of materials which comprise low-level radioactive waste and why such materials require special handling and care; and reasonably detailed explanations of alternative disposal methods and their probable effects.

§ 7. The department of environmental conservation, the department of health and the New York state energy research and development authority shall submit annually, as a part of their budget request, an estimate of the anticipated costs to carry out the purposes of this act.

§ 8. That part, entitled "LOW-LEVEL RADIOACTIVE WASTE SITING PROGRAM", of section one of chapter fifty of the laws of nineteen hundred eighty-six, (STATE OPERATIONS BUDGET), is amended to read as follows:

STATE OPERATIONS-MISCELLANEOUS

LOW-LEVEL RADIOACTIVE WASTE SITING PROGRAM

General Fund - State Purposes Account

For allocation to state departments, agencies and authorities for services and expenses including contractual services and siting and licensing of [temporary and permanent] low-level radioactive waste [storage] *disposal* facilities consistent with the Federal Low-Level Radioactive Waste Policy Act of 1980 *and Low-Level Radioactive Waste Policy Amendments of 1985* pursuant to a chapter of the laws of 1986 3,500,000

SCHEDULE

Energy Research and Development Authority	*250,000*
Department of Environmental Conservation	*1,520,000*
Department of Health	*100,000*
Commission for siting low-level radioactive waste disposal facilities	*1,630,000*
Total of Schedule	*3,500,000*

§ 9. Nothing in this act shall be construed to preclude any person from establishing or operating, in accordance with all applicable provisions of state, federal, and local law, any facilities for the temporary storage or permanent disposal of low-level radioactive waste, whether generated within or outside of the state of New York, which are not authorized or required by the provisions of this act.

§ 10. The provisions of this act shall be severable and, if any clause, sentence, paragraph, subdivision, or other part of this act shall be adjudged by any court of competent jurisdiction to be invalid, such judgment shall not affect, impair, or invalidate the remainder of this act, but shall be confined in its operation to the clause, sentence, paragraph, subdivision, or other part of this act directly involved in and adjudged invalid in the controversy in which such judgment has been rendered.

§ 11. This act shall take effect immediately.

CHAPTER 674

(See FISCAL NOTE at end of Chapter.)

AN ACT to amend the retirement and social security law, in relation to retirement of members of the division of law enforcement in the department of environmental conservation

Became a law July 26, 1986, with the approval of the Governor.
Passed by a majority vote, three-fifths being present.

The People of the State of New York, represented in Senate and Assembly, do enact as follows:

Section 1. Subdivision eleven of section three hundred two of the retirement and social security law is amended by adding a new paragraph f to read as follows:

f. Service as a sworn police officer of the division of law enforcement in the department of environmental conservation.

§ 2. Subdivision d of section three hundred seventy-five-f of such law, as added by chapter three hundred seventy-one of the laws of nineteen hundred sixty-nine, is amended to read as follows:

d. In addition to the retirement allowance provided pursuant to the plans set forth in sections three hundred eighty-three [and], three hundred eighty-three-a *and three hundred eighty-three-b* of this chapter, a member of either such plan who retires on or after April first, nineteen hundred sixty-nine with more than twenty-five years of total service shall be entitled to receive, in addition to the benefits provided pursuant to either such section and notwithstanding the limitations of

EXPLANATION—Matter in *italics* is new; matter in brackets [ ] is old law to be omitted.

Appendix I

Tables of Exclusionary and Preference Criteria

TABLE I.1 Exclusionary Criteria and the Steps of the New York State Siting Process in Which Each Criterion Was Applied

Exclusionary Criteria	Siting Factor	Regulatory Basis	SES	CAI		PSI		
				Exclusionary Screening	Comparative Analysis	GIS Screening	Map Assessments	Rescreening
Criterion 4 (existing and new mine methods)—Exclude all abandoned mines and all geologic units that are less than or equal to 30 meters below the ground surface.	Geology	**6 NYCRR 382.35(b)(1)** An underground mined repository must be located at a depth below the ground surface of greater than 30 meters.		X	X			X
Criterion 6 (all disposal methods)—Exclude all reforestation areas or parts thereof established pursuant to Article 9, Title 5 of ECL.	Natural resources	**6 NYCRR 382.21(h)(1)(vi)** The site must not be located on . . . property which is (vi) reforestation area or part thereof established pursuant to Article 9, Title 5 of the ECL; **6 NYCRR 382.21(c)(1)** The site must not be located in an area where past, present, or future exploration or exploitation of natural resources could adversely affect the land disposal facility's ability to meet the performance objectives of Section 382.10-382.15 of this Part. **10 CFR 61.50(a)(4)** Areas must be avoided having known natural resources which, if exploited, would result in failure to meet the performance objectives of Subpart C of this part.			X[a]	X		X

Exclusionary Criteria	Siting Factor	Regulatory Basis	SES	CAI		PSI		
				Exclusionary Screening	Comparative Analysis	GIS Screening	Map Assessments	Rescreening
		6 NYCRR 382.21(h)(1)(i) The site must not be located on real property owned in fee by the federal, state, or municipal governments, or in which such governments have a lesser interest, where the alienation or use of such property is restricted by constitutional provision or statute, including, but not limited to, property which is: (i) protected by Article 14 of the Constitution of the State of New York;						
Criterion 11 (all disposal methods)—Exclude all areas immediately above the Long Island Aquifer, any primary public water supply aquifer, or a principal aquifer designated by the Departments of Environmental Conservation and Health.	Ground water hydrology	**6 NYCRR 382.21(d)(3)** The site must not be located where potential adverse effects on groundwater could result in contravention of groundwater quality standards or an impairment of the best intended usage as specified in Parts 701, 702, and 703 of this title. **6 NYCRR 382.22(b)(2)** The site must not be located immediately above the Long Island aquifer; any primary public water supply aquifer; or a principal aquifer designated by the department.	X	X		X		X
Criterion 15 (aboveground and belowground methods)—Exclude all significant perennial surface water bodies.	Surface water hydrology	**6 NYCRR 382.22(a)(1)** The site must not be in a 100-year floodplain, coastal hazard area or wetland, as defined in Executive Order 11988, "Floodplain Management Guidelines," 42 Fed. Reg. 26951 (1977).		X		X		X

Table I.1, continued

Exclusionary Criteria	Siting Factor	Regulatory Basis	SES	CAI Exclusionary Screening	CAI Comparative Analysis	PSI GIS Screening	PSI Map Assessments	PSI Rescreening
		6 NYCRR 382.22(a)(2) The disposal site must not be located in a freshwater wetland designated pursuant to ECL article 24; a tidal wetland designated pursuant to ECL article 25; a coastal erosion hazard area identified pursuant to ECL article 34; or a flood hazard area as determined by the Federal Emergency Management Agency [(FEMA)] pursuant to 42 U.S.C. sections 4001, et seq.						
		6 NYCRR 382.23(b) The underground mined repository's surface facilities must not be placed in a 100-year floodplain, coastal erosion hazard area, or wetland, as defined in Executive Order 11988, "Floodplain Management Guidelines," 42 Fed. Reg. 26951 (1977).						
Criterion 16 (all disposal methods)—Exclude 100-year floodplains as defined in Executive Order 11988; coastal erosion hazard areas pursuant to ECL Article 34; or flood hazard areas as determined by FEMA pursuant to 42 U.S.C. Sections 4001, et seq.	Surface water hydrology	**6 NYCRR 382.22(a)(1)** The disposal site must not be in a 100-year floodplain, coastal hazard area or wetland, as defined in Executive Order 11988, "Floodplain Management Guidelines," 42 Fed. Reg. 26951 (1977).				X		X
		10 CFR 61.50(a)(5) Waste disposal shall not take place in a 100-year floodplain, coastal high-hazard area or wetland, as defined in Executive Order 11988, "Floodplain Management Guidelines."						

Exclusionary Criteria	Siting Factor	Regulatory Basis	SES	CAI		PSI		
				Exclusionary Screening	Comparative Analysis	GIS Screening	Map Assessments	Rescreening
		6 NYCRR 382.22(a)(2) The disposal site must not be located in a freshwater wetland designated pursuant to ECL article 24; a tidal wetland designated pursuant to ECL article 25, a coastal erosion hazard area identified pursuant to ECL article 34; or a flood hazard area as determined by the Federal Emergency Management Agency pursuant to 42 U.S.C. Sections 4001, et seq.						
		6 NYCRR 382.23(b) The underground mined repository's surface facilities must not be placed in a 100-year floodplain, coastal erosion hazard area, or wetland, as defined in Executive Order 11988, "Floodplain Management Guidelines," 42 Fed. Reg. 26951 (1977).						
		10 CFR 61.50(a)(5) Waste disposal shall not take place in a 100-year flood plain, coastal high-hazard area or wetland, as defined in Executive Order 11988, "Floodplain Management Guidelines."						
Criterion 17 (aboveground and belowground methods)—Exclude wetlands as defined in Executive Order 11988; freshwater wetlands designated pursuant to ECL Article 24; tidal wetlands designated pursuant to ECL Article 25.	Surface water hydrology	**6 NYCRR 382.21(h)(1)(ix)** The site must not be located on . . . property which is: (ix) wetlands acquired or restored in part or in whole with state moneys pursuant to ECL, Article 51, or 52, or other statutory provision;			X[b]	X		X

Table I.1, continued

Exclusionary Criteria	Siting Factor	Regulatory Basis	SES	CAI		PSI		
				Exclusionary Screening	Comparative Analysis	GIS Screening	Map Assessments	Rescreening
		6 NYCRR 382.22(a)(1) The disposal site must not be in a 100-year floodplain, coastal hazard area or wetland, as defined in Executive Order 11988, "Floodplain Management Guidelines," 42 Fed. Reg. 26951 (1977).						
		6 NYCRR 382.22(a)(2) The disposal site must not be located in a freshwater wetland designated pursuant to ECL article 24; a tidal wetland designated pursuant to ECL article 25, a coastal erosion hazard area identified pursuant to ECL article 34, or a flood hazard area as determined by the Federal Emergency Management Agency pursuant to 42 U.S.C. Sections 4001, et seq.						
		6 NYCRR 382.23(b) The underground mined repository's surface facilities must not be placed in a 100-year floodplain, coastal erosion hazard area, or wetland, as defined in Executive Order 11988, "Floodplain Management Guidelines," 42 Fed. Reg. 26951 (1977).						
		10 CFR 61.50(a)(5). Waste disposal shall not take place in a 100-year flood plain, coastal high-hazard area or wetland, as defined in Executive Order 11988, "Floodplain Management Guidelines."						

Exclusionary Criteria	Siting Factor	Regulatory Basis	SES	CAI		PSI		
				Exclusionary Screening	Comparative Analysis	GIS Screening	Map Assessments	Rescreening
Criterion 26 (all disposal methods)—Exclude areas that are within pollutant nonattainment areas.	Air quality	**40 CFR 81.333, National Ambient Air Quality Standards; DEC Air Guide 8, Map 1.** Site activities may not violate current air quality standards.		X		X		X
Criterion 28 (all disposal methods)—Exclude all areas where threatened or endangered species or designated critical habitats for such terrestrial or aquatic species or species of special concern, as determined pursuant to Title 16, Chapter 35 of the United States Code; Article 11 of the ECL; or 6 NYCRR Part 182, are present.	Ecology	**6 NYCRR 382.21(h)(2)(vii)** The site must not be located in the critical habitat of any endangered or threatened species as determined pursuant to Title 16 of Chapter 35 of the United States Code or in an essential habitat, as determined by the department, of a species which is designated as endangered or threatened pursuant to Article 11 of the ECL or designated as a species of special concern in Part 182 of this Title. An essential habitat does not include incidental areas passed in migration, other areas of casual use, or potential habitat.				X		X
Criterion 32 (all disposal methods)—Exclude all villages, towns, cities, or unincorporated places, as defined in the 1980 decennial census or more recent census of the United States, New York State, or any political subdivision thereof performed by the U.S. Census, that have an average population density of more than 1,000 persons per square mile.	Demographic patterns	**6 NYCRR 382.21(f)(4)** No part of the site shall be located within any village, town, city, or unincorporated area having an average population density of more than 1,000 individuals per square mile as determined from the results of the 1980 decennial count of the U.S. Census or more recent census of the United States, New York State, or any political subdivision thereof.	X	X		X		X

Table I.1, continued

Exclusionary Criteria	Siting Factor	Regulatory Basis	SES	CAI: Exclusionary Screening	CAI: Comparative Analysis	PSI: GIS Screening	PSI: Map Assessments	PSI: Rescreening
Criterion 36 (all disposal methods)—Exclude all lands protected by the Federal government, including . . .	Land use	**6 NYCRR 382.21(h)(1)(ii)** The site must not be located on . . . property which is:	X	X		X		X
- National Wildlife Refuge System		(ii) all or part of any national wildlife refuge established pursuant to the National Wildlife Refuge System, 16 U.S.C. Sections 668dd and 668ee;						
- Fish restoration areas		**6 NYCRR 382.21(h)(1)(iv)** The site must not be located on . . . property which is: (iv) acquired with the assistance of federal funding pursuant to the Dingell-Johnson Fish Restoration Act, 16 U.S.C. Sections 777-777i and 777k or the Pittman-Robertson Act, 16 U.S.C. Section 669-669i;				X		X
- Migratory bird reservations		**6 NYCRR 382.21(h)(1)(v)** The site must not be located on . . . property which is: (v) all or any part of any migratory bird reservation established pursuant to the Migratory Bird Conservation Act, 16 U.S.C. section 715;				X		X
- The National Wilderness Preservation System		**6 NYCRR 382.21(h)(1)(viii)** The site must not be located on . . . property which is: (viii) part of the National Wilderness Preservation System pursuant to Title 16, Chapter 23 of the U.S.C.;	X	X		X		X

Exclusionary Criteria	Siting Factor	Regulatory Basis	SES	CAI		PSI		
				Exclusionary Screening	Comparative Analysis	GIS Screening	Map Assessments	Rescreening
- National Wild and Scenic Rivers System		**6 NYCRR 382.21(h)(2)(ii)** The disposal site must not be located on . . . property which is: (ii) subject to the State Wild, Scenic and Recreational Rivers System pursuant to ECL Article 15, Title 27; or the National Wild and Scenic Rivers System, pursuant to the Wild and Scenic Rivers Act, 16 U.S.C. Section 1271 et seq.		X		X		X
- National Park System		**6 NYCRR 382.21(h)(4)** No part of the site may be located on real property which is within any national or state park.	X	X		X		X
Criterion 38 (all disposal methods)—Exclude all lands protected by New York State, including . . . - Adirondack Park - Catskill Park	Land Use	**6 NYCRR 382.21(h)(1)(i)** The site must not be located on real property owned in fee by the federal, state, or municipal governments, or in which such governments have a lesser interest, where the alienation or use of such property is restricted by constitutional provision or statute, including, but not limited to, property which is: (i) protected by Article 14 of the Constitution of the State of New York; **6 NYCRR 382.21(h)(3)** The site must not be located on real property which is within the Adirondack or Catskill Parks, as defined in ECL Ch.9-0101(1) and (2), respectively.	X	X		X		X

Table I.1, continued

Exclusionary Criteria	Siting Factor	Regulatory Basis	SES	CAI		PSI		
				Exclusionary Screening	Comparative Analysis	GIS Screening	Map Assessments	Rescreening
- Wildlife management areas		**6 NYCRR 382.21(h)(1)(iii)** The site must not be located on . . . property which is: (iii) a state wildlife management area, game refuge, fish hatchery, game farm, or any part of these or property in which the state has an interest under the fishing access or other public access program established pursuant to Article 11 or Article 13 of the ECL;		X		X		X
- Game refuges						X		X
- Game farms						X		X
- Fish hatcheries						X		X
- Boat launches						X		X
- Fish restoration areas		**6 NYCRR 382.21(h)(1)(iv)** The site must not be located on . . . property which is: (iv) acquired with the assistance of federal funding pursuant to the Dingell-Johnson Fish Restoration Act, 16 U.S.C. Sections 777-777i and 777k or the Pittman-Robertson Act, 16 U.S.C. Section 669-669I;				X		X
- New York State Wild, Scenic, and Recreational Rivers System		**6 NYCRR 382.21(h)(2)(ii)** The disposal site must not be located on . . . property which is: (ii) subject to the State Wild, Scenic and Recreational Rivers System pursuant to ECL Article 15, Title 27; or the National Wild and Scenic Rivers System, pursuant to the Wild and Scenic Rivers Act, 16 U.S.C. Section 1271 et seq.;		X		X		X

Exclusionary Criteria	Siting Factor	Regulatory Basis	SES	CAI		PSI		
				Exclusionary Screening	Comparative Analysis	GIS Screening	Map Assessments	Rescreening
- State Park System		**6 NYCRR 382.21(h)(4)** No part of the site may be located on real property which is within any national or state park.	X	X		X		X
- Municipal parks established as of 31 December 1987		**6 NYCRR 382.21(h)(5)** No part of the site may be located on real property which is within any municipal park established as of December 31, 1987 or within any addition thereto.						X
Criterion 40 (all disposal methods)—Exclude all real property that cannot be acquired for facility use by the State of New York.	Land use	**6 NYCRR 382.21(f)(2)** The site must be located on real property to which the State of New York can obtain title in fee or any interest therein, as may be necessary.						X
		6 NYCRR 382.21(f)(6) The site must not be located on any lands or reservations of Indian tribes or nations or on lands that are not subject to the laws and regulations of the State of New York, including but not limited to, lands owned by the federal government.						
Criterion 41 (all disposal methods)—Exclude all Indian reservations and lands under jurisdiction of Indian nations.	Land use	**6 NYCRR 382.21(f)(6)** The site must not be located on any lands or reservations of Indian tribes or nations or on lands that are not subject to the laws and regulations of the State of New York, including but not limited to, lands owned by the federal government.	X	X		X		X

Table I.1, continued

Exclusionary Criteria	Siting Factor	Regulatory Basis	SES	CAI: Exclusionary Screening	CAI: Comparative Analysis	PSI: GIS Screening	PSI: Map Assessments	PSI: Rescreening
Criterion 44 (all disposal methods)—Exclude all lands in mineral soil groups 1-4, as designated by the New York State Land Classification System, that are in active agricultural production.	Land use	**6 NYCRR 382.21(h)(6)** The site must not contain more than 5 acres of lands in active agricultural use in mineral soil groups 1-4 as classified by the New York State Land Classification System, established by 1 NYCRR Part 370.			X[b]	X	X	X
Criterion 46 (all disposal methods)—Exclude all lands within the boundaries of the Western New York Nuclear Services Center in West Valley, New York.	Land use	**Ch. 673, Art. 29, Title 3, Sec. 29-0303, Subdiv. 7 (as amended by Chapter 913)** Such site or sites shall not include the western New York nuclear service center.		X				X
		6 NYCRR 382.21(a)(1) The site must not be located at the Western New York Nuclear Service Center in West Valley, New York.						
Criterion 57 (all disposal methods)—Exclude all area that are listed, nominated, or eligible for listing in the National or State Registers of Historic Places.	Cultural resources and aesthetics	**6 NYCRR 382.21(h)(2)(i)** The disposal site must not be located on real property where the alienation or use of such property is restricted by federal or state statute, including, but not limited to, property which is: (i) listed, nominated or eligible for listing in the National Register of Historic Places pursuant to the National Historic Preservation Act, 16 U.S.C. Sections 470 et seq., or the State Register of Historic Places pursuant to Article 14 of the Parks, Recreation and Historic Preservation Law;			X[a]	X		X

Exclusionary Criteria	Siting Factor	Regulatory Basis	SES	CAI		PSI		
				Exclusionary Screening	Comparative Analysis	GIS Screening	Map Assessments	Rescreening
Criterion 58 (all disposal methods)—Exclude all lands that have been dedicated to or acquired for the purpose of being dedicated to the State Nature and Historical Preserve.	Cultural resources and aesthetics	**6 NYCRR 382.21(h)(1)(vii)** The site must not be located on . . . property which is: (vii) dedicated to or acquired for the purpose of being dedicated to the State Nature and Historical Preserve pursuant to ECL Article 45, State Nature and Historical Preserve Act;						X

NOTE: CAI = Candidate Area Identification; CAIR = *Candidate Area Identification Report*; CFR = *Code of Federal Regulations*; DEC = New York State Department of Environmental Conservation; FEMA = Federal Emergency Management Agency; GIS = Geographic Information System; NYCRR = *New York Code of Rules and Regulations*; PSI = Potential Sites Identification; U.S.C. = *United States Code*.

[a]The Siting Commission's *Excluded Areas Report* states that this criterion was first applied during PSI.
[b]The CAIR states that a 1977 Soil Conservation Service map of prime farmlands was used as a proxy for Criterion 44 during the Comparative Assessment stage of CAI.

TABLE I.2 Preference Criteria and the Steps of the New York State Siting Process in Which Each Criterion Was Applied

			CAI		PSI			
Criterion Number	Siting Factor	Criterion Statement	Initial Screening	Comparative Analysis	GIS Screening	Map Assessments	Rescreening	Weight[a]
1	Geology	Prefer areas that contain uniformly distributed soils, sediments, and/or rock that are relatively undeformed and devoid of fractures, faults, and other discontinuities that may influence predictions of performance of the facility.		X	X		X	45
2	Geology	Prefer areas of lower seismic hazard based on predicted maximum horizontal ground acceleration as given by Algermissen et al. (1982).	X		X		X	20
3	Geology	Prefer areas that do not demonstrate significant past or active subsurface dissolution.		X	X		X	35
5	Geology	Prefer host geologic units that can accommodate the vertical and horizontal space required for all disposal facilities and any necessary buffer zone.	X	X	X		X	35
7	Natural resources	Prefer areas that are distant from active or nearby abandoned mines.			X		X	20
8	Natural resources	Prefer areas that do not have likely economic mineral resource potential.			X		X	20
9	Natural resources	Prefer existing mines and nearby areas that do not have economic mineral or energy resource potential.						
10	Natural resources	Prefer areas that are distant from active or abandoned oil or gas fields, underground injection wells, underground gas storage area, and areas of solution mining.	X		X		X	20
12	Ground water hydrology	Prefer areas that are distant from primary or principal aquifers, hydrogeologic units that fit the definition of principal aquifers, and well head areas for community water supply systems.	X	X	X		X	55
13	Ground water hydrology	Prefer areas that are distant from significant surface water features that are likely to be sustained by groundwater discharge through the site.	X		X		X	40
14	Ground water hydrology	Prefer unconsolidated stratigraphic units that restrict groundwater flow and retard radionuclide movement.	X		X		X	40

Criterion Number	Siting Factor	Criterion Statement	CAI		PSI			Weight[a]
			Initial Screening	Comparative Analysis	GIS Screening	Map Assessments	Rescreening	
18	Surface water hydrology	Prefer areas that are distant from wetlands as defined in Executive Order 11988; fresh water or tidal wetlands designated pursuant to ECL Articles 24 or 25; or areas that may be under Federal jurisdiction as wetlands. The latter includes areas that have any of the following conditions (1) Hydric soils (SCS or U.S. Army Corps of Engineers classification), (2) National Wetlands Inventory classification, and (3) extensive areas of less than 2 percent slope.		X	X		X	30
19	Surface water hydrology	Prefer sites that are well drained and free from areas of potential ephemeral flooding or ponding from streams and runoff.		X		X	X	30
20	Surface water hydrology	Prefer sites exhibiting no existing or potential erosional characteristics that could adversely affect waste containment.		X		X	X	35
21	Surface water hydrology	Prefer areas where there is no potential for flooding from failure of upstream man-made impoundments.		X			X	25
22	Surface water hydrology	Prefer areas where location of a disposal facility is unlikely to impair the best usage of surface waters, as measured against water quality standards.	X	X	X		X	30
23	Meteorology and climatology	Prefer areas with relatively low annual amounts of precipitation.	X		X		X	20
24	Meteorology and climatology	Prefer areas that are not prone to chronic severe weather phenomena such as heavy snowfall.	X		X		X	15
25	Meteorology and climatology	Prefer sites that are not prone to frequent incidence of severe weather.			X		X	20
27	Air quality	Prefer areas where a significant PSD increment is available for consumption, especially for total suspended particulates and inhalable particulates.					X	30

Table I.2, continued

Criterion Number	Siting Factor	Criterion Statement	CAI		PSI			Weight[a]
			Initial Screening	Comparative Analysis	GIS Screening	Map Assessments	Rescreening	
29	Ecology	Prefer areas that are distant from critical habitats of terrestrial or aquatic species that are threatened or endangered or that have been proposed for inclusion as threatened or endangered species, as determined pursuant to Title 16, Chapter 35 of the United States Code, Article 11 of the ECL, or 6 NYCRR Part 182; and areas inhabited by other important species.			X		X	55
30	Incompatible nearby activities	Prefer areas that are as far as practicable from any existing facility or natural source that could generate and/or release radioactive or non-radioactive materials that may mask the environmental monitoring program.		X	X		X	20
31	Incompatible nearby activities	Prefer areas that are distant from nearby activities that could interfere with the ability of the facility to meet the performance objectives of 6 NYCRR Part 382, Subpart C.				X	X	20
33	Demographic patterns	Prefer areas that exhibit low population densities.	X		X		X	45
34	Demographic patterns	Prefer areas that are distant from highly populated places, as defined by the U.S. Census (greater than 2,500 persons).		X	X		X	45
35	Demographic patterns	Prefer areas that are distant from concentrations of nonresident population.				X	X	25
37	Land use	Prefer areas that are distant from lands protected by the Federal government.	X		X		X	10
39	Land use	Prefer areas that are distant from lands protected by New York State.	X		X		X	10
42	Land use	Prefer areas that are distant from Indian lands.	X		X		X	10
43	Land use	Prefer government-owned lands that have not been otherwise excluded from consideration.					X	20
45	Land use	Prefer areas where development or significant local population increases will not be imminent.					X	20

Criterion Number	Siting Factor	Criterion Statement	CAI Initial Screening	CAI Comparative Analysis	PSI GIS Screening	PSI Map Assessments	PSI Rescreening	Weight[a]
47	Transportation	Prefer areas near the interstate highway system, other major (multilane, limited access) roadways, or existing rail lines.		X	X		X	10
48	Transportation	Prefer sites served by routes that are not currently congested.					X	10
49	Transportation	Prefer sites that utilize routes that do not require new construction or structural improvements to handle waste shipments.				X	X	5
50	Transportation	Prefer sites where access is available by more than one mode.			X		X	5
51	Transportation	Prefer areas that are close to major waste generators.	X		X		X	20
52	Transportation	Prefer sites for which the access routes exhibit low accident rates, as measured by the actual numbers of accidents, resulting property damage, and lives lost.					X	15
53	Transportation	Prefer sites served by routes that do not pass through incorporated areas upon leaving the major transportation system (e.g., interstate highway system).			X		X	10
54	Socioeconomics/ community services	Prefer sites where there is a sufficient work force available with the required labor skills to construct and operate the disposal facility.					X	10
55	Socioeconomics/ community services	Prefer sites with an adequate or excess housing stock (in terms of quality, number, and affordability) within a reasonable commuting distance or time.					X	10
56	Socioeconomics/ community services	Prefer sites, within reasonable commuting distance or time, where municipal services (e.g., sewer, water, schools, emergency services) have not reached capacity and where the disposal facility will not create a disproportionate demand on these services.					X	15
59	Cultural resources and aesthetics	Prefer sites that do not contain other archeological, cultural, or historical resources.		X	X		X	20

Table I.2, continued

Criterion Number	Siting Factor	Criterion Statement	CAI		PSI			Weight[a]
			Initial Screening	Comparative Analysis	GIS Screening	Map Assessments	Rescreening	
60	Cultural resources and aesthetics	Prefer sites where the disposal facility will not degrade the present scenery or viewshed.					X	15
61	Cultural resources and aesthetics	Prefer sites that do not have people or animals in close proximity that may be sensitive to noise.					X	10

NOTE: CAI = Candidate Area Identification; ECL = New York State *Environmental Conservation Law*; GIS = Geographic Information System; NYCRR = *New York Code of Rules and Regulations*; PSD = Prevention of Significant Deterioration; PSI = Potential Sites Identification; SCS = Soil Conservation Service

[a]Criterion weights listed apply to aboveground or belowground disposal methods.

Appendix J

Quality Assurance Documents

American Society of Mechanical Engineers. 1986. Quality Assurance Program Requirements for Nuclear Facilities, Supplement 17S-1 (NQA-1).

New York State Department of Environmental Conservation. 1987. Regulations for Low-Level Radioactive Waste Disposal Facilities (Certification of Proposed Sites and Disposal Methods). New York Code of Rules and Regulations Title 6, Part 382. Albany, New York.

New York State Legislature. 1986. Low-Level Radioactive Waste Management Act. Laws of the State of New York, Chapter 673 (July 26):2801-2817.

U.S. Nuclear Regulatory Commission. 1983. Final Technical Position on Documentation of Computer Codes for High-Level Waste Management (NUREG-0856). Washington, D.C.

U.S. Nuclear Regulatory Commission. 1988. Peer Review for High-Level Nuclear Waste Repositories: Generic Technical Position (NUREG 1297). Washington, D.C.

U.S. Nuclear Regulatory Commission. 1988. Qualification of Existing Data for High-Level Nuclear Waste Repositories (NUREG 1298). Washington, D.C.

U.S. Nuclear Regulatory Commission. 1988. Standard Format and Content of a License Application for a Low-Level Radioactive Waste Disposal Facility—Safety Analysis Report. Rev. 1 (NUREG-1199). Washington, D.C.

U.S. Nuclear Regulatory Commission. 1989. Quality Assurance Guidance for Low-Level Radioactive Waste Disposal Facility (NUREG-1293). Washington, D.C.

Appendix K

Examples of Windshield Survey Forms

SHEET___ OF___

WINDSHIELD SURVEY CHECKLIST -- PLANNING CRITERIA

PREPARED BY ____________________ DATE ______ APPROVED BY _______
CHECKED BY ____________________ DATE ______ APPROVED BY _______

CANDIDATE AREA ____

POTENTIAL SITE(S)_________________________ T ___ R ___ SEC._______

FIRST IMPRESSIONS

Are there any obvious unfavorable conditions found upon arrival at the site that would clearly prohibit development of the waste disposal facility? If no, proceed with remainder of checklist. If yes, state the prohibiting conditions below and proceed to the next site. ____
____
____
____
____.

CRITERIA NOT PREVIOUSLY ADDRESSED

CRITERION 31
INCORPORATE NEARBY ACTIVITIES (ACTIVITIES) CHK. BY _____

Identify any nearby activities that could interfere with the ability of the facility to meet the performance objectives of 6 NYCRR Part 382, Subpart C. Such facilities could include areas of irrigated agriculture (groundwater withdrawal) mining and quarrying areas, industrial incinerators, transmission lines or pipelines.

Comment:____
____
____
____
____.

CRITERION 32

POPULATION DENSITY (EXCLUDE > 1,000) CHK. BY _____

Delineate the boundaries of unincorporated areas and other concentrations of populations not already evaluated. Identify the number of residential units within. Multiply the number of units by 2.5 to arrive at approximate total population.

Comment:____
____
____
____
____.

CRITERION 35
NONRESIDENT POPULATION

Identify the locations, names, and types of facilities that are located in or near the site that would account for nonresidents inhabiting the area. This would include recreation facilities (parks, campgrounds, trails, picnic areas), manufacturing plants, office buildings, schools, and other institutional properties. Attempt to estimate size/number of visitors by building size, approximate number of parking spaces, etc.

Comment:____
____
____
____
____.

CRITERION 38
STATE-PROTECTED LANDS (MUNICIPAL PARKS) CHK. BY _____

Identify the locations, names, and types of municipal parks located in or near the potential site. Note creation date if available.

Comment:____
____
____
____
____.

CRITERION 40
REAL PROPERTY NOT ABLE TO BE ACQUIRED CHK. BY _____

Identify the locations, names, and types of any properties that do not appear to be acquirable for the facility. The exact nature of these facilities has not been identified, but should include cemeteries or institution-owned lands (school district forests as an example found in some states). Ownership other than private should be noted.

Comment:____
____
____
____
____.

CRITERION 44
SOIL GROUPS 1-4 CHK. BY _____

This criterion has been evaluated in the PSIR, however, issues related to it remain. Data on farming activity and crop types will be important in settling these issues as well as input to the preliminary performance assessments. Where possible, delineate areas of active agriculture (remember pasture/grazing) and indicate crop type. The latter will be difficult; it may be possible to differentiate by general classifications (i.e., cropland and pasture; orchards, groves, nurseries, etc.; confined feeding operations; rangeland, and other.

Comment:____
____
____
____
____.

CRITERION 43
OTHER GOVERNMENT-OWNED LANDS CHK. BY _____

See Criterion 40. Note government-owned lands not already analyzed. Municipal solid waste landfills, transfer stations or sewage treatment facilities outside municipal boundaries are potential examples.

Comment:____
____
____
____
____.

CRITERION 45
POPULATION GROWTH CHK. BY _____

Note the locations, size and type of development that may be developed in the near future. Clearing and road building are good signs of future/present development. Sometimes a billboard or sign is placed along a roadway announcing future projects.

New county or town roads (or even in some cases improvements) may signal the direction, albeit not quantity, of growth.

Comment:____

____

____

____

____.

CRITERION 48
CONGESTION CHK. BY _____

This will be difficult to analyze on a site-specific basis. However, it may be noticable through travels to candidate areas that contain segments of major highway systems are extremely busy. Note any such segments, as well as any site-specific access roads that appear to be overloaded.

Comment:____

____

____

____

____.

CRITERION 49
NEW CONSTRUCTION/IMPROVEMENTS CHK. BY _____

As most sites are located in very rural settings and are accessible only by town roads, note any segments that are narrow, in need or repair (surface breaking up), or any old or unmaintained or weight-restricted bridges crossed. Give the highway system in the immediate vicinity a straightforward rating of either good, fair, or poor.

Comment:____

____

____

____

____.

CRITERION 52
ACCIDENT RATES CHK. BY _____

This will most likely not be able to be evaluated, however, if any guard rails appear to be damaged from vehicular contact, please note. Give highway system in immediate area an observed rating of good, fair, or poor. Take into consideration speed restrictions, changes in slope, number and severity of curves, etc.

Comment:____
____
____
____
____.

CRITERION 54
LABOR FORCE CHK. BY _____

Note local employment opportunities such as types of manufacturing, does the sites's perceived labor shed/community area include a relatively large city, is the local economy dominated by agriculture, etc.

Comment:____
____
____
____
____.

CRITERION 55
HOUSING STOCK CHK. BY _____

While passing through cities, villages, hamlets and unincorporated places proximate to the site, notice the availability of residential properties being marketed. Note what appear to be anomalies, e.g., many or no houses for sale. This will be a qualitative evaluation. (The number of employees at the facility during construction will be approximately 100 and 30 during operation.)

Comment:____
____
____
____
____.

CRITERION 56
INFRASTRUCTURE CHK. BY _____

If while reading local newspapers, etc., an article specifies a particular need for additions to the school or hospital, more emergency personnel and equipment, a new highway, water or sewer system, note it here. Don't anticipate the ability to visualize infrastructure insufficiencies.

Comment:____
____
____
____
____.

CRITERION 60
VIEWSHEDS CHK. BY _____

Evaluate the site with respect to its elevation, vegetative cover and the overall ability to permit the facility to be viewed by the public. Delineate viewsheds where possible.

Comment:____
____
____
____
____.

CRITERION 61
NOISE CHK. BY _____

Identify the locations of nearest residences or recreation facilities, etc., that may be within earshot of the facility. Consult with biologists with regard to animal species within or nearby the site. If able to approximate furthest distance noise could be heard, do so.

Comment:____
____
____
____
____.

CRITERIA COVERED IN PREVIOUS SCREENING STEPS

Comment on conditions you see relating to the following criteria that have been used in GIS screening.

CRITERION 6
REFORESTATION AREAS **CHK. BY** _____

Comment: ____
____
____
____.

CRITERION 30
INCOMPATIBLE NEARBY ACTIVITIES (SOURCES) **CHK. BY** _____

Comment: ____
____
____
____.

CRITERION 32
POPULATION DENSITY (EXCLUDE > 1,000) CHK. BY _____

Comment: ____
____
____
____.

CRITERION 33
POPULATION DENSITY (PREFER LOW DENSITY) CHK. BY _____

Comment: ____
____
____
____.

CRITERION 34
HIGHLY POPULATED PLACES CHK. BY _____

Comment: ____
____
____
____.

CRITERION 36/37
FEDERAL LANDS CHK. BY _____

Comment: ____
____
____
____.

CRITERION 38/39
NEW YORK STATE LANDS CHK. BY _____

Comment: ____
____
____
____.

CRITERION 41/42
INDIAN LANDS CHK. BY _____

Comment: ____
____
____
____.

CRITERION 44
SOIL GROUPS 1-4 CHK. BY _____

Comment: ____
____
____
____.

CRITERION 46
WEST VALLEY FACILITY CHK. BY _____

Comment: ____
____
____
____.

CRITERION 47
PROXIMITY TO MAJOR HIGHWAYS CHK. BY _____

Comment: ____
____
____
____.

CRITERION 50
MULTIMODAL ACCESS CHK. BY _____

Comment: ____
____
____
____.

CRITERION 51
PROXIMITY TO MAJOR WASTE GENERATORS CHK. BY _____

Comment: ____
____
____
____.

CRITERION 53
ROUTES THROUGH INCORPORATED PLACES CHK. BY _____

Comment: ____
____
____
____.

CRITERION 57
HISTORIC PLACES CHK. BY ______

Comment: ____
____
____
____.

CRITERION 58
ARCHEOLOGICAL SITES CHK. BY ______

Comment: ____
____
____
____.

RFW432

SHEET___ OF___

WINDSHIELD SURVEY CHECKLIST -- GEOSCIENCE CRITERIA

PREPARED BY ____________________ DATE ______ APPROVED BY _______
CHECKED BY ____________________ DATE ______ APPROVED BY _______

CANDIDATE AREA ____

POTENTIAL SITE____________________________ T ___ R ___ SEC._______

FIRST IMPRESSIONS

Are there any obvious unfavorable conditions found upon arrival at the site that would clearly prohibit development of the waste disposal facility? If no, proceed with remainder of checklist. If yes, state the prohibiting conditions below and proceed to the next site. ____
____
____
____
____.

CRITERIA NOT PREVIOUSLY ADDRESSED

CRITERION 19
DRAINAGE CHARACTERISTICS CHK. BY _____

Provide written descriptions of observations relating to how well-drained the site is, and comment on the potential for ephemeral flooding of ponding due to streams or runoff. Evaluate road cuts for soil permeability. Evaluate surface for drainage network development. Check culverts, check dams, levees, etc., for potential flooding. Provide a good, fair, or poor rating to the site. Identify data needs that could better aid your decision.

Comment:____
____
____
____
____.

CRITERION 20
POTENTIAL OR EXISTING EROSIONAL CHARACTERISTICS CHK. BY _____

Provide written descriptions of observations relating to existing or potential erosional characteristics (i.e., slope failures, mass wasting, sheet erosion, etc.) and comment on the potential for adverse impacts on waste containment. Evaluate soil strength in road cuts. Use aerial photography or maps to assess post or potential landslides. Note springs or seeps if possible. Provide a good, fair, or poor evaluation.

Comment:____
____
____
____
____.

CRITERION 21
UPSTREAM IMPOUNDMENTS CHK. BY _____

Provide written descriptions of observations relating to the potential for ephemeral flooding from failure of upstream man-made impoundments. Include potential impact to power lines, access roads, etc. Provide a good, fair, or poor rating for the site.

Comment:____
____
____
____
____.

OTHER FACTORS/ISSUES

SLOPE

Provide written description of topography in terms of percent-area by percent slope. Describe slope type (i.e., exposed rock, colluvium, glaciogenic deposits) and condition (i.e., vegetated, forested, dissected, etc.). Comment on slope for constructability issues. Note any slope protection measures currently in use, e.g., bolts, nets, walls, etc. Provide a good, fair, or poor rating to the site.

Comment:____
____
____
____
____.

CONSTRUCTABILITY

Provide written description of conditions that may result in issues impacting the constructability (time, money, politics) of an option at the site(s) (i.e., remoteness, access roads, power, existing developments/land use, etc.). Evaluate potential for intrusion. Provide a good, fair, or poor rating.

Comment:____
____
____
____
____.

FLAWS

Comment on any obvious or potential flaws that the site(s) exhibits that would restrict development or eliminate the site(s) from further consideration. Ask yourself if it was you task to build a facility here what would the potential drawbacks be.

Comment:____
____
____
____
____.

CRITERIA COVERED IN PREVIOUS SCREENING STEPS

Comment on the conditions you see relating to the following criteria that have been used in GIS screening.

CRITERION 1
STRATIGRAPHIC COMPLEXITY CHK. BY _____

Aboveground/Belowground Disposal Methods

Comment: ____
____
____
____.

Mine Repository Disposal Method Only

Comment: ____
____
____
____.

CRITERION 3
SUBSURFACE DISSOLUTION CHK. BY _____

Comment: ____
____
____
____.

CRITERION 5
THICKNESS AND AERIAL EXTENT OF GEOLOGIC UNIT CHK. BY _____

Aboveground/Belowground Disposal Methods

Comment: ____
____
____
____.

Mine Repository Disposal Method Only

Comment: ____
____
____
____.

CRITERION 7
NEARBY MINES CHK. BY _____

Comment: ____
____
____
____.

CRITERION 8
RESOURCE POTENTIAL CHK. BY ____

Do your observations confirm the favorability ranking in the draft report? ____yes ____no; if no, comment below with observation and suggested modification (attach sheets if necessary).

Comment: ____
____
____
____.

CRITERION 10
OIL AND GAS FIELDS CHK. BY _____

Comment: ____
____
____
____.

CRITERION 12
AQUIFERS CHK. BY _____

Comment: ____
____
____
____.

CRITERION 13
GROUNDWATER DISCHARGE ZONES CHK. BY _____

Comment: ____
____
____
____.

CRITERION 14
RADIONUCLIDE RETARDATION (PERMEABILITY) CHK. BY _____

Comment: ____
____
____
____.

CRITERION 16
FEMA 100-YR FLOODPLAINS CHK. BY _____

Comment: ____
____
____
____.

RFW429

SHEET___ OF___

WINDSHIELD SURVEY CHECKLIST -- WETLANDS AND ECOLOGY CRITERIA

PREPARED BY ____________________ DATE ______ APPROVED BY _______
CHECKED BY ____________________ DATE ______ APPROVED BY _______

CANDIDATE AREA ____

POTENTIAL SITE(S)__________________________ T ___ R ___ SEC._______

____

FIRST IMPRESSIONS

Are there any obvious unfavorable conditions found upon arrival at the site that would clearly prohibit development of the waste disposal facility? If no, proceed with remainder of checklist. If yes, state the prohibiting conditions below and proceed to the next site. ____
____
____
____
____.

CRITERIA NOT PREVIOUSLY ADDRESSED

Criteria relating to wetlands and ecology have been evaluated in previous screening steps. However, additional analyses pertaining to these criteria are to be conducted during the windshield surveys as stated below.

CRITERIA 17 and 18
WETLANDS

Provide written summaries on the occurrence of the following subcriteria for each site:

- Wetland habitats (refer to CAIR and DEC maps)
- Streams/floodplains
- Hydric soils
- 0 - 1% slopes

    o Natural habitat islands or peninsulas within agricultural fields

- Application of above to immediate downgradient areas

Comments: ____

CRITERIA 28 AND 29
ECOLOGY

Provide written summaries on the occurrence of the following subcriteria for each site:

- o Proximity to Threatened/Endangered species records in screening report
- o Occurrence on site of habitats of T/E species based upon DEC list and species ranges
- o General habitat descriptions: habitats of "Important" species based upon experience

- Application of above to immediate downgradient areas

Comments: ____

RFW454

Reconnaissance Summary   Site Number__________________

PLANNING CRITERIA

Overall Rating (New criteria only)	1 (Least Favorable)	3 (Favorable)	5 (Most Favorable)
Criterion Number	Rating		
31	1	3	5
32	1	3	5
35	1	3	5
38	1	3	5
40	1	3	5
44	1	3	5
43	1	3	5
45	1	3	5
48	1	3	5
49	1	3	5
52	1	3	5
54	1	3	5
55	1	3	5
56	1	3	5
60	1	3	5
61	1	3	5

Comments:

Reconnaissance Summary Site Number________________

GEOLOGY CRITERIA

Overall Rating (New criteria only)	1 (Least Favorable)	3 (Favorable)	5 (Most Favorable)
Criterion Number	Rating		
19	1	3	5
20	1	3	5
21	1	3	5
Other Issues	Rating		
Slope	1	3	5
Constructability	1	3	5
Flaws	1	3	5

Comments:

Reconnaissance Summary Site Number__________________

WETLANDS/ECOLOGY CRITERIA

Overall Rating	1 (Least Favorable)	3 (Favorable)	5 (Most Favorable)
Criterion Number	Rating		
17	1	3	5
18	1	3	5
28	1	3	5
29	1	3	5

Comments:

RFW443